GYANENDRA ENUMERATIO PLANTAMEDICA

INDIAN MEDICAL SCIENCE SERIES NO. 15

Gyanendra Enumeratio Plantamedica

GYĀNENDRA AUṢADHĪYA PĀDAPĀVALĪ

DR. GYANENDRA PANDEY

Edited by
SUMITRA PANDEY
College of Science
Gurukal Kangri University
HARDWAR, U.P., INDIA

SRI SATGURU PUBLICATIONS
A Division of
INDIAN BOOKS CENTRE,
DELHI, INDIA

Published by :
SRI SATGURU PUBLICATIONS
Indological and Oriental Publishers
A Division of
INDIAN BOOKS CENTRE
40/5, Shakti Nagar,
Delhi-110007
(INDIA)

FIRST EDITION : DELHI, 1993

ISBN 81-7030—353-2

Token to mark fourth Contributory Decade outset of Dr. Gyanendra Pandey. Inheritance of pursuit to herbal knowledge in Family Tradition of Prof. Ram Nath (1905-1991). Dedicated to the Nation, Science and Humanity in Transmission to further Generations.

Printed at :
Kiran Mudran Kendra
A-38/2 Maya Puri, Phase-I,
New Delhi-110064

PRINTED IN INDIA

CONTENTS

PREAMBLE

Theme and Plan :

Enumeration of medicinal plants of India and adjacent regions especially plant drugs used in indigenous systems and traditions of medicin, in the form of glossary and checklist for making a guide catalogue as reference handbook in precised manner for ready consultation of drugs and their botanical sources alongwith families (possibly incorporating name changes wherever needed, and available).

Pattern of Presentation :

Principal Name (or Relevent name) of Drug
(Roman English)

Botanical Name	*Family*
(chief or major prevalent plants or source plants	(chief plant)

Relevent plants substitutes—Adulterants
claimed plant sources and also
allied species)

LIST OF FAMILIES

Acanthaceae
Agaricaceae
Amaranthaceae
Amaryllidaceae
Alangiaceae
Apocynaceae
Annonaceae
Apiaceae
Aristolochiaeeae
Araceae
Anacardiaceae
Araliaceae

Berberidaceae
Betulaceae
Bixaceae
Bignoniaceae
Bombacaceae
Burseraceae
Basellaceae
Balsminaceae

Ceratophyllaceae
Caesalpiniaceae
Capparidaceae
Celastraceae
Caricaceae
Chenopodiaceae
Cannaceae
Cruciferae
Cuppresaceae
Cucurbitaceae
Cycadaceae
Cyperaceae
Convolvulaceae
Casuarinaceae
Cannabinaceae
Caryophyllaceae
Cleomaceae

Dilleniaceae
Dioscoreaceae
Dipteracarpaceae

Ericaceae
Elaeocarpaceae
Euphorbiaceae
Erythroxylaceae

Fabaceae
Ficioidaceae
Flacourtiaceae
Fumariaceae
Fagaceae

Gentianaceae
Gnetaceae
Guttifereae

Hamamelidaceae
Hippocastanaceae
Hypocreales
Hypericaceae

Iridaceae
Juglandaceae

Lamlaceae
Lauraceae
Lecythiadaceae
Liliaceae
Loranthaceae
Lythraceae
Leeaceae

Magnoliaceae
Malvaceae
Meliaceae
Malpighiaceae
Menispermaceae
Moringaceae
Moraceae
Musaceae
Myristicaceae
Myrtaceae
Mimosaceae
Marsileaceae

Nymphacaceae
Nyctaginaceae

Oleaceae
Onagraceae
Orehidaceae
Oxalidaceae

Primulaceae
Palmae
Pandanaceae
Papaveraceae
Parmeliaceae
Pedaliaceae
Poaceae
Polygonaceae
Plantaginaceae
Plumbaginaceae
Punicaceae
Pinaceae
Passifloraceae
Portulacaceae
Platanaceae

Ranunculaceae
Rhizocarpae
Rhamnaceae
Rubiaceae
Rosaceae
Rutaceae

Salicaceae
Samydaceae
Santalaceae
Sapindaceac
Salvadoraceae
Saxifragaceae
Sellginallaceae
Scrophulariaceae
Simarubaceae
Salicaceae
Symplocaceae
Smilacaceae
Solanceae
Sterculiaceae

Theaceae
Tamaricaceae
Taxaceae
Thymelaceae
Tiliaceae
Typhaceae

Ulmaceae
Urticaceae

Valerianaceae
Verbenaceae
Violaceae
Vitaceae
Zannichelliaceae
Zingiberaceae
Zygophyllaceae

Compositeae	:	Asteraceae
Papilionaceae	:	Fabaceae
Gramineae	:	Poaceae
Umbellifercae	:	Apiaceae
Labiateae	:	Lamiaceae
Cruciferae	:	Brassicaceae
Aizoaceae	:	Ficoidaceae
Capparidaceae	:	Capparaceae
Cupuliferae	:	Fagaceae

SECTION I

APARĀJITĀ—GIRIKARNIKA
Clitoria ternatea Linn. Fabaceae

ARANYAKULATTHA
Cassia absus Linn. Caesalpiniaceae

ARJAKA
Orthosiphon pallidus Royle. Lamiaceae
Orthosiphon aristatus (Bl.) Miq.
=*O. stamineus* Benth.

AŚMANTAKA
Ficus cordifolia Roxli. Moraceae
Bauhinia malabarica Roxli.

ASTHISAMHARA
Cissus quadrangula L. Vitaceae
Vttis quandrangularis wall ex ξ. & A.

AHIPHENA
Papaver somoniferum Linn Papaveraceae
=*P. setigerum* Dc.

ATIBALĀ
Abutilon indicnm (Lamk.) SW. Malvaceae
=*A. indicum* Var. *populifolium* (Lamk.) W. & A.
Abutilon indicum (L) Sweet. ssp.
guineense (Schumach) Borssum.
=*A asiaticum* (L.) Sweet.
=*A. histum* (Lamk.) Sweet.
=*A. graveolens* var. *hirtum* (Lamk.) Masters
A. muticum (DC) Sweet.
=*A. glaucum* (cav.) Sweet.

AMLIKA
Tamarindus indica Linn. Caesalpiniaceae

ARIMEDA—IRIMEDA

Acacia leucophloea Willd Mimoseae

A. farnesiana Willd

ARISTAKA

Sapindus emarginatus Vahl. Sapindaceae

=*Sapindus trifoliatus* Vahl. var. *emarginatus* (Vahl.) Cooke

S. trifoliatus L.

S. laurifoliatuṣ auct. non L.

S. mukorossi Gaertn.

Sapindus rarak Dc.

=*Dittelasma rarak* (Dc) Hiern

Blighia sapida Kon.

=*S. obovatuṣ* W. & A.

ARKA Asclepiadaceae

Calotropis procera (Ait.) Ait. f. ssp. *hamiltohü*

=C. prowra auct. mon (Ait.) Ait. f. (w.) Ait.

C. procera (Ait.) R Br.

C. gigantea (Linn.) R. Br. ex Ait.

ARJUNA

Terminalia arjuna W. & A. Comleretaceae

ALABU—IKSVAKU Cucurlitaceae

Lagenaria siceraria (Molina) Standley

=L. leucantha (Duchesne) Rusby

AŚOKA

Saraca indica Linn. Caesalpiniaceae

Saraca asoca (Roxli) De wilde

=*S. indica* sensu Baker

Saraca declinata (Jack) Miq.

=*S. cauliflora* Baker.

Polyalthia longifolia Benth. & Hook" f.

AMLAVETASA—VṚKṢĀMLA

Garcinia pedunculata Roxb. Guttifereae

Garcinia atroviridis Griff

G. gunumi-gatta (L) Rob.

=*G. cambogia* (Gaertn) Desr.
G. camlogia var. *conicarpa* (W) Anderson.
G. Kydia Roxb.
=*G. kydiana* Roxb. ex w.
=G *wallichü* choisy
=*G. Cowa* sensu T. And. p.p. non Roxb.
G. acuminata Planchon & Triana
=G. morella sensu T. and.
G Pietoria Roxb. sensu T. And.
G. morella Desr.
G. xanthochymuș Hook. f. & T. And
=*G. pictorius* (Roxb.) A. Arcy. non *G. pictoria* Roxb.
G. cornea Linn.
G. cowa Roxb.
G. dulcis Kurz.
G. heterandra wall.
G. indica chois
G. mangostana Linn.

Rheum australe D. Don. Polygonaceae
=*R. emodi* wall ex Meisn.
Rheun delavayi Franch.
=*R. nepalense* Kitamura
R. acuminatum Royle
R. nobila Hook. of & Th.
K. officinale Bailen
REVANDACĪNĪ *R. officinale* Bailon
R. palmatum Linn.
R. spiciforme Royle
R. wellianum Royle

AŚVABALĀ
Meliolotus indica (L) All.
=*M. parviflora* Desf.
Indigofera trifoliate Linn.
Indigofera prostrats willd.
=*I. trifoliata* auct. non L.
Medicago sativa L ssp. *falcata* (L) Naeg$_3$ & Thall.
=*M. falcata* L.

ASANA—BĪJAKA

Pterocarpus marsupium Roxb. Fabaceae

Terminalia alata Heyne ex Roth
T. tomentosa (Roxb.) W. & A.

ATASĪ

Linum usitatissimum Linn. Linaceae

Linum corymbulosum Reichmeh.
=*L, strictum* var. *corymbulosum* (R) Plench.
L. perenna L.
=*L. perenne* var. *stocksianuum* (Boiss) Boiss.

ATIMUKTAKA—MADHAVI

=*Hiptage benghalensis* (L) Kurz. Malpighiaceae
=*H. parviflora* W. & A.
=*H. madablota* Gaertn.

ATIVIṢĀ—PRATIVISA

Aconitum heterophyllum wall. Ravuxculaeeae

Aconitum bisma (Ham.) Rapaics.
=*Caltha bisma* Ham.
A. palmatum D. Don.

APAMARGA

Achyranthes aspera Linn. Amaranthaceae
A bidentata Blune.

Achylauthes porphyristachya wall
=*A. aspera porphyristachya* (wall) Hook. f.

AJAKARṆA Dipterocarpaceae

Dipterocarpus indicus Bedd.
=*Dipterocarpus turbinatus* Gaertn.

AŚVAKARNA Dipterocarpaceae

Dipteracarpus alatus Roxb.
=*D. incanus* Roxb.

D. costatus Gaertn. f.
=D. alatus sensu. Dyor b. p., non Roxb.
D. gracilis Bl.

=*D. pilosus* Roxb.
D. obtusisfolius Teysm.
=*D. vestitus* wall.
D. grandiflorus (Blxo) Bico.
=*D. ptery gocalyx Scheff.*

SARJAKA

Vateria indica Linn. Dipteracarpaceae
Shorea disticha (Thw.) Ashton
=*Vateria disticha* Thw.
Vatica disticha A. Dc.

AŚVASANDHĀ Solanaceae

Withania somnifera Dunal.
(incl. *W. Ashwagandha* Kaul.)
W. Cougulans Dunal.

AŚVATTHA—PIPPALA.

Ficus religiosa Linn. Moraceae

AJAMODA Apiaceae

Trachyspermum roxburghianuna (Dc) Craili.
=*Carum roxburghianum* Kurz.
Apium glavedens Linn.

ERAṆḌA

Ricinus communis Linn. Euphorbiaceae

ERVĀRUKA Cucurbitaceae

Cucumis utillissimus Roxli.

ELĀ

Elettaria cardamonum Maton. Zingiheracae
Amomum subulatum Roxb.
Amomum xanthophleblum Baker
A. aromaticum Roxb.
A. xanthioides wall.
A. biflorum Jaeq.
A. pterocarpum Thw.

ELAVĀUKA

Prunus cerasus Linn. Rosaceae
P. avium Linn.

ERAKĀ–GUNDRĀ

Typha angustata Bory & chauli Typhaceae

T. elephantina Roxli

T. laxmanny Tiypha angustifolia L. Lepch.

=*T. Javanica* S. ex R.

ERAṆḌAKARAṬĪ

Carica papaya Linn. Caricaceae

AKṢOṬA

Juglans ragia var. *Kumaonia* C. Dc. Juglándaceae

=*J. ragia* Hook. fi.

AGASTYA

Sesbania grandiflora (Linn.) Pers. Fabaceae

AGURU—AGARU

Aquilaria malaccensis Lamk. Thymalaceae

=*A. agallocha* Roxli.

A. khasiana Hallier.

=*A. officinalis* Cl.

AGNIMANTHA

Clerodendrum multifolorum...... Verbenaceae

=*C. phlomidis* Linn. f, *Clerodendron phlomidis* Limm. f.

Premna corymbosa Rottl.

=*P. integrifolia* L.

ANKOLA-ANKOTA (Alangiaceae)

Alañgium salvifolium (L. k.) wangerin ssp.

salvifolium Mukerjee,

=*A. salvifolium* L. f. *A. lamarckü* Thwaites

Alangium alpinum (Cl.) W.W Smith & Cave

=*Marlea alpina* (Cl.) Gamble ex Brandis

A. barbatum Baillon

=*M. barbata* R. Br. excl.

A. chinensis (Lour) Harms.

=*M. begoniaefolia* Roxb.

***A. salvifolium* ssp. *hexapetalum* (Lamk.) Wangerin.**

=*A. lamarckü* sensu Cl. incl. var. *glandulosum* (Thw.) Cl.

AMBAṢṬHAKĪ

Hibiscus sabdariffa Linn. Malvaceae

ARKAPUṢPĪ

Holestemma ada-kodien J.A. Schultes. Asclepiadaceae
=*Sarcostemma annulare* Roth.
=*Holostemna annulare* (Roth.) Schum.
=*H.* rheedü wall.
=*H. rheedianum* Spreng.

AMṚTAPHALA

Pyrus communis Linn. Rosaceae

ADHOPUṢPĪ

Trichodesma indicum var. *amplexicaule* (Roth.) cook. Boraginaceae
=*T. amplexicaule* Roth.
Trichodesma sedgwickianum Banerjee.
=*T. amplexieaule* sensu cl.
=*T. indicum* var amplexicaule Cooke.

ARAṆYAJĪRAKA

Centratherum anthelmintica (willd,) Kuntze. Asteraceae
=*Conyza anthelmintica* L.
=*Veronia anthelmintica* (L.) willd.

ANTAMŪLA

Tylophora indica (Burm. f.) Merrill. Asclepiadaceae
=*T. asthmatica* (L. f.) W.&A.
Tylophora fasciculata Ham ex Wight.
T. tenuissima (Roxb) W&A ex W.
=*T. tenuis* Blume.
T. multiflora (W.&A. ex W.) Alston.
=*T. iphisio* Decne.

ĀḌHAKI

Cajanus cajan (Linn.) Huth. Fabaceae
=*C. obcordifolia* Singh

ĀTMAGUPTA—KAPIKACCHU

Mucuna pruriens (L.) Dc. Fabaceae
=*M. prurita* Hook.
M. prurita Hook.
=*M. pruriens* Bak..

M. capitata W.&A.
M. gigantea Dc.
M. monosperma
M. nigricans (Lour) Steudel
=*M. imbricata* Dc.

ĀMALAKĪ—DHĀTRĪ

Emblica officinalis Linn. Euphorbiaceae
=*Phyllanthes emblica* Linn.

ĀMRA

Mangifera indica Linn. Anacardiaceae

ĀMRĀTAKA

Spondias pinnata (L. f.) Kurz. Anacardiaceae
=*Spondias mangifera* willd.
S. macrophylla wall.

Spondiäs indica (W.&A.) Airy Shaw
=*Solenocarpus indicus* W.&A.

ĀRAGVADHA

Cassia fistula Linn. Caesalpiniaceae

ĀRUKA

Prunus domestica L. var. insititia (L.) Bailey. Rosaceae
=P. communis Huds. var. insititia (L.) Hook. f.
P. domestica Linn
=*P. communis* Huds.

Prunus armeniaca Linn.

P. persica Batach.

ĀRDRAKA—ṢUNṬHĪ

Zingiber officinale Roscoe Zingiberaceae

ĀLUKA

Dioscorea belophylla Voigt ex Haines Dioscoreaceae
=*D. glabra* auct. non Roxli
D. pentaphylla L.
=D. jacquemontü Hook f.
Dioscorealata L.
—*D. rubella* Roxli
D. prazeri Prain & Burkill
—*D. clarkei* Prain & Burkill
—*D. sikkimenses* P.&B.
D. deltoidea wall

ĀVARTAKĪ

Cassia auriculata Linn Caesalpiniaceae

ĀMRAGANDHIHARIDRĀ

Curcuma amada Roxli. Zingiberaceae

AVARTANĪ

Helicteres isora Linn. Sterculiaceae

ĀKĀṢAVALLĪ

Cuscata reflexā Roxli. Convolvulaceae
—*C. santopaui* Banerji
Cuscuta chinensis Lam.
C. hyalina Roth.
Cassytha filiformis Linn.

ĀLUKĪ (RAKT. LUBHEDA)

Colocasia esculenta (L.) Schott. Araceae
=*Arum esculentum* L.
=*C. antiquorum* Schott.

ĀKARAKARABHA

Anacyclus pyrethrum Dc. Asteraceae
=*Anthemis pyrethrum* Linn.

Spilanthes clava Dc.
=*S. acmella* var. *clava* (Dc.) cl.
=*S.* acmella auct. non (L.) Murr.
=*S. pseudo-acmella* auct. non (L.) Murr.
Spilanthes oleracea L.
=*S. acmella* var. *oleracea* cl.
Spilanthes paniculata Dc.
=*S.* acmella auct. non (L.) Murr.
Spilanthes sinensis (Pers.) Ames.
=*S. sinensis* ssp.
australis (R. Br.) Lindley.
=*S. sinensis* var. *amaena* (M. Bieles) Hara.

ĀKĀŚANIMBA

Millingtonia hortensis Linn. f. — Bignoniaeeae

IKṢU

Saccharum officinarum Linn. — Poaceae

IKSURAKA—KOKILĀKṢA

Hygrophila auriculata (Schumach) Heine. — Acanthaceae
=*Astercantha longifolia* (L) Nees.
=*Hygrophila spinosa* T. Ander.
Matham Arah :— *Hygrophila salicifolia* Nees.
—*H. angustifolia* auct. non R. Br.
H. phlomoides Nees.

INGUDI

Balanites aegyptiaca (L.) Delile. — Simarubaceae
=*Xinenia aegyptiaca* L.
=Balanites roxburghü Planch.

INDRAVĀRUṆĪ

Citrullus colocynthis Schrad. — Cucurbitaceae
Bislambhi =*Cucumis trigonus* Roxle.
=C. pseudo-colocynthis Royle.
Mahakala (Makal)=*Trichosanthes tricuspidata* Lour.

=T. palmata Roxb.
T. lepiniana (Naudin) Cogn.
T. bracteata (Lamk.) Voigt.

ĪṢADGOLA (AŚVAKARṆA)

Plantago major L. var. *angusta* angusta (Pilgar) Yamazake Plantaginaceae
=*P.* asiatica' var. *angusta* Pilger.
P. erosa wall
=*P. major* auct. non L.
P. ovata Forsk.

P. amplexicaulis cav. ssp.
bauphula (Edge.) Rech. f.
=*P. bauphula* Edge.
=*P. amplexicaulis* sensu Hook. f.
P. lanceolata Linn.
P. himalaica Pilger.
=*P. brachyphylla* Edge ex Decne.
P. ciliata Desf.
P. tibetica Hook f.
P. depressa willd.
P. psyllium L.

UDUMBARA

Ficus racemosa Linn. Moraceae
=*F. glomerata* Roxb.

UDDĀLAKA

Kodrava =*Paspalum scrobiculatum* Linn. Poaceae
=*P. commersonü* Lamk
=*P. cartilagineum* Presl.
Oḍār =*Sterculia villosa* Roxb.

UPAKUNCIKA

Nigella sativa Linn. Ranunculaceae

UPODIKĀ

Basella alba L. var *rubra* (L.) Stewart Basellaceae

=*B. rubra* L.
B. alba L.
=B. rubra sensu Hook. f., p.p. non L.

URUMĀṆA

Khubani *Prunus armeniaca* Linn. Rosaceae

UŚĪRA

Vetiveria zizanioides. (Linn.) Nash. Poaceae
=*Andropogon muricatus* Ketz.

UṢṬRAKAṆṬAKA

Echinops echinatus Roxb. Asteraceae
Echinops niveus wall ex Royle
E. cornigerous Dc.

KADARA

Acacia polycantha willd.
=*Acacia suma* (Roxb.) Kurz.
=*Mimosa suma* Roxb.

KAṬUKĀ

Picrorhiza kurroa Royle ex Benth. Scrophulariaceae.
Picrorhiza scrophularüflora Pennel.
=P. kurroa auct. non Royle.
Hellebores miger Linn.
H. viridis Linn.

KAṆṬAKĀRI

Solanum surattense Burm. f. Solanaceae

KṢUDRĀ

=S. xanthocarpum Seh. & wendl.
=S. mecani Sant.

BṚHATĪ

Solanum indicum L.
Solanum ferox L.
=*S. indicuum* L. nom. ambig.

Solanum anquivü Lamk.
S. indicum auct. non L.
Solanum myriacanthum Dunal.
=*S. khasianum* clarke.
S. viarum Dunal.
=*S. khasianum* cl. var. Chatterjeeanum Gupta.
S. torvum Sw.

KATAKA

Strychnos potatorum Linn. Ranunculaceae

KADALĪ

Musa paradisiaca Linn. Musaceae
M. paradisiaca Linn. var. sapientum Kuntze.
Musa balbisiana colla.
=*M. sapientum* auct. non. L.
=*M. sikkimensis* Kurz.
=*M. hookeri* King.

KARKAṬĪ

Cucumis utilissimus Roxb. Cucurbitaceae
Cucumis cucudain Linn.

KAPIṬTHA

Feronia acidissima L. Rutaceae
=*F. limonia* (L.) Swingle.
=*F. elephantum* Correa.

KAMALA

Nelumbo nucifera Gaertn. Nymphacaceae
=*Nelumbium speciosum* willd.

KAMPILLAKA

Mallotus philippinensis Muell.-Arg. Euphorbiaceae

KARAMARDA

Carissa congesia w. Apocynaceae
=*C. carandus* L. var. congesta (W.) Bedd.

Carissa carandus Linn.
carissa spinarum Linn.

KARAVIRA

Nerium indieum Mill. Apocynaceae
=*N. odorum* Solander.
Nerium olander Linn.
Pita Kaner —*Caseabcia thevetia* (L) Lippoid
Pita =Thevetia peruviana (Pers.) K. Schun
T. nerüfolia Juss. ex Steud.

KARĪRA

Capparis decidua (Forsk.) Edgew. Capparidaceae
=*C. aphylla* Roth.

KĀKODUMBARIKĀ

Ficus hispida Linn. Moraceae

KARKAṬAṢṚNGĪ

Pistacia chinensis Bunge ssp *integerrima* (Stewart) Rech. f. Anacardiaceae
=*P. integerrima* Stewart.
Pistacia khinjuk Stocks.
=*P. integerrima* auct. non Stewart.
=*Rhus karkatasingee* Royle.
Rhus succedanea L. var. *succedanea.*
=*R. suceedanea* var. *himalaica* Hook. f.
=*R. succedanea* var. *sikkimensis* Hook. f.
=*R. succedenea* var. *acuminata* (Dc.) Hook.

KARKANDHU

Zizyphus nummularia W.&H. Rhamnaceae

KARKOṬAKA

Momordica diocia Roxb. Cucurbitaceae

KARCURA

Curcuma zedoria Rose. Zingiberaceae

KARPŪRA

Cinnamomum camphora Nees. & Eberm. Lauraceae
Dryobalanops aromatica Gaertn. f.
=*D. camphora* Colebr.

KALAMBA

Ipomoea reptans (Linn.) Poir. Convolvulaceae

KĀKANĀSA

Martynia annua L. Pedaliaceae
=*M. diandra* Glox.
Pentatropis capensis (L. f.) Bullock.
=*Cynanchum capense* L. f
=*P. microphylla* (Roxb.) W.&A.

KĀLĀJĀJĪ

Nigella sativa Linn. Ranunculaceae

KĀKĀDANĪ

Cardiospermum halicacabum Linn. Sapindaceae

KĀKAMĀCĪ

Solanum nigrum Linn. Solanaceae

KĀKOLĪ

Roscoea procera Wall.
R. purpurea Smith.
=*R. purpurea* var. *exilis* sensu. Baker.
=*R. purpurea* var. *procera* (Wall.) Baker.

KĀRAVĪ

Carum carvi Linn. Apiaceae
C. bulbocastanum Koek.

KĀRAVELLAKA

Momordica charantia Linn. Cucurbitaceae

KĀRPĀSA

Gossypium barbadens Linn.
G. arboreum L var. *obtusifolium* (Roxb) Roberty.
=*G. herbaceum* var. *obtusifolium* (Roxb.) Masters
G. hirsutum L. var. *hirsutum* Borssum.
=*G. herbaceum* var. *hirsutum* (L.) Masters.

KĀRPASĪ

Hibiscus cancellatus Roxb. Malvaceae
Thespesia lampas Dalz. & Gibs.

KĀLĪYAKA

Coscinium fenestratum (Gaertn.) Colebr.
Menispermaceae

KĀṢA

Saccharum spontaneum Linn. Poaceae

KĀṢMĀRRĪ—GAMBHĀRĪ

Gmelina arborea Linn. Verbenaceae
Gmelina asiatica Linn.
G. elliptica Smith.
=*G. villosa* Roxb.
G. phillippensis Cham.
=*G. hystrix* Kurz.

KASAMARDA

Cassia occidentalis Linn. Caesalpiniaceae

KIṆIHĪ

Albizzia procera (Roxb) Benth. Mimosaceae
=*Albizia procera* (Roxb.) Benth.

KIRĀTATIKTA

Swertia chirata Buch-Ham. Gantianaceae
S. chiraylta (Rexb ex Fleming) Karsten.
=*S. chirata* (Wall.) Cl.
=*S. tongluensis* Burkill.

Swertia angustifolia Ham. ex D. Don.
var. *pulchella* Burkilt.
=*S. pulchella* cl.
=*S. affinis* cl.
S. alata Royle ex D. Don.
S. angustifolia Buch.-Ham.
S. corymbosa Wight.
S. densifolia (Griseli) Kashyappa
=S. decussata Nimmo ex Graham.
S. cilliata (G. Don.) Burtt.
=*S. purpurascens* cl.
S. lawü Burkill.
S. paniculata Wall.
S. perennis Linn.
Centaurium centaurioides (Roxb) Raof Hemadri.
=*Erythraea roxburghü* D. Don.
=*E. roxburghii* var *raxatilis* Cl.
Exacum tetragonum Roxb.
E. bicolor Roxb.
E. pedunculatum Linn.

KARPŪRAHARIDRĀ (AMRAGANDHA)
Curcuma amada Roxb.
Curcuma caesia Roxb.

KOLAKANDA (VANAPALĀṆḌU)
Drimia indica (Roxb.) Jessop Liliaceae.
=*Urginea indica* (Roxb.) Kunth.
=*Scilla indiea* Roxb.
=*U. coromandeliana* (Roxle.) Hook. F.
=*U. wightiana* Hook. F.
=*U. gobindappae* Boraiah & Fatima.
Scilla hyacinthina (Roth.) McBride.
=*Ledebouria hyacinthina* Roth.
=*S. indica* (W.) Baker.

KAIDARYA (MISṬNIMBA)

Murraya koenigü Spreng. Rutaceae
Murraya paniculata (L.) Jack.
M. exotica Linn.

KĀKAJAṄGHĀ

Peristrophe paniculata (Forsk.) Brummitt.
=*P. bicalyculata* (Retz.) Nees. Acanthaceae
=*Dianthera bicalyculata* Retz.
Peristrophe paniculata var.
subaequibracteata (Bennet.) Bennet
=*P. bicalyculata* var. *ruleacquibracteata* Bennet.
Leea acquata L.
=*L. hirta* Roxb. ex Hornem.
=*L. kurzū* Cl.

KUṄKUMA

Crocos sativus Linn. Iridaceae

KUṬAJA

SITA : *Holarrhena antidysenterica* Wall. Apocynaceae
ASITA : *Wrightia tinctoria* R. Br.
W. arborea (Dennst.) Mabberley.
=*W. tomentosa* R. & S.

KUMUDA—KUMUDINI

Nymphae alba Linn. Nymphacaceae
Nymphaea rulera Roxb. ex Andrews.
=*N. lotus* auct. non L.
=*N. magnifica* (Salisb) Conard.
N. pubescens willd.
=*N. lotus* var. *pubescens* (willd.) Hook. F.
N. nouchali Burm. F.
=*N. lotus* auct. non. L.

KURAṆṬIKĀ

Celosia argentea Linn. fa. *cristata* (L.) Schinz. Amaranthaceae
=*C. cristata* L.

KURUVAKA

Rhododendron arboreum Smith. Ericaceae

KULTTHA

Dolichos biflorus Linn. Fabaceae
Macrotyloma uniflorum (Lamk.) Verde. var *uniflorum*
=*Dolichos uniflorus* Lamk.
=*D. biflorus* senso Murray.

KALINDA

Cucumis vulgaris Schrad. Cucurbitaceae

KUŚA

Desmostachya bipinnata Stapf. Poaceae

KUṢṬHA

Saussurea lappa C.B. Clarke. Asteraceae

KUSUMBHA—LAṬVĀKA

Carthamus tinctorius Linn. Asteraceae

KŪṢMĀṆḌA

Benincasa hispida (Thunb.) Cucurbitaceae
=*Cucurbita hispida* Thunb.
=*C. pepo* F.l.
=*Benincasa cerifera* Savi.

KṚṬAVEDHANA

Luffa acutangula (Linn.) Roxb. Cucurbitaceae
=*L. acutangula* var. amara (Roxb.) el.
=*L. umbellata* (Klein ex willd.) Roem.

KETAKA—KETAKI

Pandanus tectorius Soland ex Parkinson. Pandanaceae
Pandanus fascicularis Lamk.
=*P. tectorius* auct. non Soland.

KEBUKA

Costus speciosus (Koenig) Sm. Zingilberaceae
=*C. speciosus* var *nepalensis* (Rose.) Baker.

KODRAVA

Paspalum scrobiculatum L. Poaceae
=*P. cartilagineum* Presl.

KOLA—BADARA

Zizyphus jujuba Lam. Rhamnaceae
Z. mauritiana Lamk.
=*Z. jujuba* (L.) Gaertn.
Z. mauritiana var. *fruticosa* (Haines) Selast. & Henry
=*Z. jujuba* var. fruticosa Haines.

KOVIDĀRA—KAṄCANĀRA

Bauhinia purpurea Linn. Caesalpiniaceae
B. variegata Linn.

KOṠĀTAKĪ

Luffa acutangula (Linn.) Roxb. Cucurbitaceae
L. acutangula (L.) Roxb. *amara* clarke.
L. cylindrica (Linn.) M. Roem.
L. aegyptiaca Mill. ex Hook. F.
=*L. cylindrica auct. non* (L.) Roem.

KOṠĀMRA

Schleichera bosa (Lour.) Merr. Sapindaceae
S. obosa (Lour.) Oken
=*S. trijugo* Willd.

KAṄGU

Setaria italica Beauv. Poaceae

KAṄKOLA

Piper cubeba Linn. Piperaceae

KAṬAPHALA

Myrica esculenta Ham.
=*M. nagi* Hook. F. non Thunb.

KATṚṆ—ROHIṢA

Cymbopogon martinü (Roxb.) Wats. Poaceae
=*Andropogon martinü* Roxb.
=*C. motia* Gupta.
C. citratus (DC.) Stapf.

KADAMBA
NĪPA *Anthocephalus chinensis* (Lamk.) A. Rich. Rubiaceae

=*Cephalanthes chinensis* Lamk.

RAJKADAMA=*A. cadamba* (Roxb.) Miq.
=*A. indicus* Rich.

DHARAKADAMA
GIRIKADAAMBA—HRIDRU *Adina cordifolia* Benth. & Hook. F.
DHULIKADAMBA *Mitragyna parviflora* (Roxb.) Korth. var. *parviflora*
=*Stephegyme parviflora* (Roxb.) Korth.
M. parvifoia var. *microphylla* (Kurz.) Ridadale
=*Nauclea parvifolia* var. *microphylla* Kurz.

KARMARAṄGA

Averrhoa bilimbi Linn. Oxalidaceae
A. carambola Linn.

KALĀYA

Pisum sativum Linn. Fabaceae

KAṠERUKA

Scirpus kysoor Roxb. Cyperaceae
S. grossus Linn. F. *Cyperus esculenta* Linn.

KAKAJAṄGHA

Peristrophe paniculata (Forsk.) Brummitt. Aeanthaceae
=*Dianihera bicalyculata* Retz.
=*P. bicalyculata* (Retz.) Nees.

KUNDA

Jasminum pubescens Willd. Oleaceae

KAṆDIRA—JALADHĀNYAKA

Ranunculus seeleratus Linn. Ranunculaceae

Ranunculus aquatilis Linn.
var. *capillaceus* Dc.
R. trichophyllus Chaix
=*R. aquatilis* L. var. *trichophyllus* Hook. F. & Thoms.
R. arvensis Linn.
R. *microphyllus* Hand.
=R. *flaccidus* Hook. F. & Thoms.
R. falcatus Linn.
R. līngua Linn.
R. muricata Linn.
R. pennsylvanicus L.F. ssp. *riparius* (Edge.) Riedl.
=*R. pennasylvaticus* auct. non L.
R. adoxifolius Hand.
=*R. affinis* sensu Hook. F. & Thoms

KAṢṬHAPĀṬALA

Stereospermum colals (Dillw.) Mabberley. Bignoniaceae
=*Stereospermum chelonoides* Dc.
=*S. tetragonum* Dc.
=*S. personatum* (Hassk.) Chatterjee

Stereospermum colals var. *anqustifolium* (Haines) B. & R.
=*S. chelonoides* var. *angustifolium* Haines.
=S. *personatum* var. *angustifolium* (Haines) Santisuk.

KUMĀRĪ

Aloe barbadensis Mill. Liliaceae
=*A. vera* Tourn ex Linn.
=*A. indica* Role.
=*A. litoralis* Koening.

A. perryi Baker.
A. succotrina Lam.

KUKUNDARA

Blumea lacera (Burm. F.) Dc. Asteraceae
=*Conyza lacera* Burm. F.
=*Blumea subeapitata* Dc.
Blumea lacera var. *blumei* De.
=*B. laeiniata* (Roxb.) Dc.
=*B. sinuata* (Lour.) Merrill.
B. balsami fera Dc.
B. densiflora Dc.
B. eriantha Dc.
B. lanceolaria (Roxb.) Dc.
=*B. myriocephala* Dc.
=*B. spectabilis* var. *spectalilis* (Dc.)
=*B. spectabilis* var. *spectabilis* (Dc.)
B. lanceolaria var. *spectabilis* (De.) Hook. f.
=B. *spectabilis* Randeria Dc.
B. riparia (Bl.) Dc.
=*B. chinensis* Dc.

KUBJAKA

Rosa moschata Herrm. Rosaceac
Rosa brunonü Lindley.
=*R. moschata* auct. non Mill.
=*R. moschata* var. *nepalensis* Lindl.

KJĀŪTA (KA)

Anacardium occidentale Linn. Anacardiaceae

KĪṬAMĀRĪYAVĀNĪ

Artemisia maritima L. var. *willdenoviana* Bess. Asteraceae
=*A. maritima* var. *maritima* fa. *rubicaule* Badhwar.
=*A. kurramensis* Quazilbash.
Artemisia absinthum Linn.
A. biennis roilld.
=*A. tournefortiana* Reichb.
A. persica Boiss.
A. gmelinü web ex Stechm.
=*A. sacrorum* Ledeb

A. capilaris Thunle.
=*A. scoparia* Waldst.
A. drancunculus Linn.
A. anuua Linn.

KṚṢṆABĪJA

Ipomoea nil (L.) Roth. — Convolvulaceae
=*I. hederacea* auct non Jaeq.

KALAMBĪ

Ipomoea aquatica Forrk. — Convolvulaceae
=*I. raptans* (L.) Pers.

KĀKATUṆḌĪ

Asclepias currasavica Līnn. — Asclepiadacee

KUMBHĪ

Careya arborea Roxle. — Myrtaceae
Planchonia andamanica King.
=*C. valida* sensu Parkinson.
Cochleospermum religiosum (L.) Alston.
=*C. gossypium* Dc.

ṚDDHI

Habenaria intermedia D. Don. — Orchidaceae
Eulophia nuda Lindl.

RṢBHAKA

Malaxis acuminata D. Don.
=*M. biloba* (Lindl.) Ames.
=*M. wallichii* (Lindl.) Dele.
=*M. wallichii* Lindl.
=*M. acuminata* var. *biloba* (Lindl.) Ames.
=*M. wallichii* var. *omphaloides* Par ex Ridl
=*M. wallichii* var. *biloba* (Lindl.) Hook. f.
Malaxis andamanica (King. & Pantl.) Balakr. & Vasud.

=*M. wallichii* auct. non Lindley
=*M. biaurita* (Lindl.) Kuntze.

ṚDDHI – VṚDDI

Habenaria intermedia D. Don. Orchidaceae
Habenaria spp.
Habenaria acuminata Thw.
Peristylis goodyeroides (D. Don) Lindley.
=*Habenaria goodyeroides* D. Don.
Diphylax griffithii (Hook. f.) Krzl.
=*Habenaria griffithii* Hook. f.

KṢAVAKA

Centipeda minima (Linn.) A. Br, et Aeschers. Asteraceae
=*A. mnima* L.
=*Centipeda orbicularis* Lour.

KṢĪRAKĀKOLĪ

Roscoea procera wall. Zingiberaceae
=*R. purpurea* Sm.
Roscoea purpurea Smith.
=*R. procer* var. *exilis* sensu Baker.
=*R. procera* var. *procera* (wall) Bakar.
=*R. brandisii* (Baker) K. Schum.
=*R. purpurea* Smith var.
Brandisii King ex Baker.
Nomocharis oxypetala Royle.
N. nana E.H.
Fritilaria roylei Hook.
F. cirrhosa D. Don.
F. imperialis Linn.
Lillium wallichianum J.A. & J.H.
Schultes var. *neligherrense* (W.) Hara.
=*L. neilgherrense* W.

KṢĪRACAMPAKA

Plumeria acuminata Ait. Apocynaceae

=*P. acutifolia* Poir.
P. rubra L.

KHADIRA

Acacia catechu Willd. Mimosaceae

KHARJŪRA—KHARJURĪ

Phoenix sylvestris Roxb. Palmae
P. dactylifer Linn.
P. humilis Royle.
P. acaulis Buch.-Ham.
P. pusilla Gaertn.

KHARBŪJA

Cucumis melo Linn. Cucurbitaceae

GAJAPIPPALĪ

Scindapsus officinalis Schott. Araceae
Piper chaba Hunter.

GAVEDHUKA

Coix lachryma jobi Linn.

GILOḌYA

Ceropegia candelabrum L. ssp *tuberosa* (Roxb.) Huber.
Asclepiadaceae
=*C. tuberosa* Roxb.
C. vincaefolia Hook.
=*C.* hirsuta var. *vincaefolia* (Hook.) Hook. f.
=*C. polyantha* Blott. & Mclann.

GUGGULU

Commiphora wightü (A.) Bhandari Burseraceae
=*Balsamodendron mukul* Hook ex Stocks.
=*B. roxburghü* Stocks.
=*Commiphora mukul* (Hook. ex stocks) Engl.
Commipnoia stocksiana (Engl.) Engl.

=*Balamea stocksiana* Engl.
=*Balsamodendron pubuescens* Stocks.
non commiphora pubescens (W. & A.) Engl.
Commiphora berryi (A) Engler
=*B. berriyi* A.
C. opobalsamum (Linn.) Engl.

BALSAM

C. myrrha Nees.
B. myrrha T. Nees

GUOUCĪ

Tinospora cordifolia (*willd.*) *Miers.* Menispermaceae
=*Tinospora crispa* (L.) Hook. f. & Thoms.
=*Menispermum crispum* L.
=*T. mastersi* Dids.
T sinensis (Lour) Merrill.
=*T. tomentosa* (colebr.) Hook. f. & Thoms.
=*T. malabarica* (Lamk.) Hook f. & Thoms.

GANGERUKĪ

Grewia tenax (Forsk.) Fiori Tiliaceae
=*G. populifolia* Vahl.
G. hirsuta Vahl.

GṚSTIKA

Dioscorea bulbifera Linn. Dioscoraceae

GOKṢURA

Tribulus terrestris Linn. Zygophyllaceae
=*Tribulns terrestris* auct. non L.
Tribulus *lanuginosus* L.
T. longipatalus Viv. ssp. *longipetalus*
=*T. alatus* Delib Ghafoor
Pedalium murax Linn.

GOJIHVĀ– GOJI

GAVJAVAN

Onosma bracteatum wall. Boraginaceae
Caccinia glauca savi

ADHOPUSPI

Trichodesmo indicum Var *amplexicaule* (Roth). *Cooke.*
=*T. amplexicaule* Roth
T. sedgwickianum Banerjee
=*T. amplexicaule* sensu Cl.
=*T. indicum* var. *amplexicaub* Cooke.

DUGDHPHENI

Taraxacum officininale Weber.
T. eripodnm Dc.
=*T. officinale* var *eriopoda* (Dc) Hook. f.

GOGIHVASAKA

Launaea asplenifolia Hook. f.
Anchusa strigosa Lobill.
Echium amoenum Fisch.
Borago officinalis Linn.
Maharanga lycopsioides (Fisch.) Johnstan.
=*Onosma lycopsioides* Fircher.
=*Lycopis arvensis* sensu Fisch. non. L.
Lycopis arvensis Linn.
Anchusa ovata Lehm.
=*Lycopis arvensis* auct. non. L.
Arnebia benthamü (wall ex G. Don.) Johnston
=*Echium benthamü*wall ex G. Don.
=*Lithospermum benthamü* (wall.) Johnston.
=*A. euchroma* (Royle.) Johnston.
=*L. euchromon* Royle.
=*M. perennis* (Schrenk.) Boiss.

GODHUMA

Triticum aestivum Linn. Poaceae

GARJAN

Daucus carota Linn. var. *sativa* Dc. Apiaceae

GUṄJĀ

Abrus precatorius Linn Fabaceae
Abrus pulchellus Wall ex. Thwi ssp. *pulchellus*

GHOṆTA—GOPAGAOṆTĀ

Zizyphus xylopyra willd. Rhamnaceae

Zizyphus glaberrima (Sedg.) Santapau
=*Z. xylopyra* var. *glaberrima* Sedg.

GIRIPARPATA

Podophyllum hexandrum Royle ex Camb. Berberidaceae
=*P. emodi* wall ex Honigberger
=*P. emodi* var. *hexandrum* (wall ex Honig) chatt. & Muker.

GANDHAMŪLA

Hibiscus esculentus Linn. Malvaceae

GRAṆṬHIPARṆA

Leonotis nepetaefolia R. Br. Lamiaceae

GORĀṆĪ

Cyamopsis tetragonoloba Taub. Fabaceah

CAKRAMARDA

Cassia tora Linn. Caesalpiniaceae
=*C. obtusifolia* Linn.

Cassia obtusifolia L.
=*C. tora* sensu Baker
C tora var. *obtusifolia* (L.) Haines.

CAṄCU

Corchorus aestuans L. Tiliaceae
=*C. acutamgulus* auct. non Forsk.
C. depressus (L.) Stocks.
=*C. antichorus* Raeusch.
C. capsularis Linn.
C fascicularis Linn.
C. olitorius Linn.
C. tridoculaaris Linn.

Hastiśuṇḍī *Heliotropium indicum* Linn.
Melochia corchorifolia Linn.

CAṆAKA

Cicer arietinum Linn. Fabaceae

CANDANA

ŚVETA *Santalum album* Linn. Santalaceae
RAKTA *Pterocarpus santalinus* Linn. f
KALEYAKA *Coscinium fenestratum* (Gaertn.) Coleb.

CAMPAKA

Michelia champaca Linn. Magnoliaceae
Michelia doltsopa Buch Ham.
=*M. Champaca* sensu Hook. f. & Thoms.
=*M. monipurensis* Watt. ex Brandis.

CAVYA

Piper chava Hunter. Piperaceae
P. officinarum Cas Dc.

CĀṄGERĪ

Oxalis corniculata Dinn. Oxalidaceae
Oxalis acetosella L. ssp. *griffithii* (Edge & Hook. f.) Hara
=*O. griffithii* Edge & Hook. f.
=*O. acetosella* Linn.
Oxalis dehradunensis Raizada.
=*O. intermedia* A. Rich.
=*O. latifolia* auct. non H.B.K.

CITRAKA

ŚVETA *Plumbago zeylanica* Linn. Plumbaginaceae
RAKTA *P. indica* L.
=*P. rosea* L.
NILA *Plumbago capensis* Thunb.

CIRABILVA

Holoptelia integrifolia Planch.

CIRBHITA (1)

Cucumis melo var. *momordiaca* Duthie & Fuller
Cucurbitaeeae

CĪNA -CĪNAKĀ

Panicum milliaceum Linn. Poaceae

CORAKA–CAṆḌĀ

Angelica glauca Edgw.
A. archanglica Linn.
=*A. archangelica* var. *himalaica* (cl.) Nasir.
=*A. officinalis* Hoofm. Apiaceae

CUKRA—CUKRIKA

Rumex vesicarlus Linn. Polgonaceae
R. acetosella Linn.

CANDRASŪRA

Lepidium sativum Linn. Cruciferae

CILHAKA

Casearia elliptica Willd.
=*Casearia tomentosa* Roxb. Samydaceae
C. zeylanica (Gaertn.) Thw.
=*C. esculenta* Roxb.

CICIṆḌA

Trichosanthes anguina Linn Cucurbtaceae

CHHATRAKA

Agaricus compestris Linn. Agaricaceae

JAMBIRA–JAMBIRA (PHALA)

Citrus limon (L.) Burm f. Rutaceae
=*C. medica* ssp. *limon* L.
=*C. medica* var. *limonium* Hook. f.

JAMBIRA (KA)

Allium prattii wright.
=*A. victorialis* var. *angustifolium* Hook. f.
=*A. victorialis* non L.
A. victorialis Linn.
A. humile Kunth.
=*A. govanianum* Wall ex Baker.
A. wallichii Kunth.
Lemongrass/*Cymbopogon ciratus* (Dc.) Stapf.

JALPIPPALI

Phyla modiflora (L.) Greene. — Verbenaceae
=*Lippia modiflora* (L.) A. Rich.

JĀTĪ

Jasminum grandiflorum L. — Oleaceae
=*J. officinale* L. var. *grandiflorum* (L.) Stokes.
=J. *officinale* fa. *grandiflorum* (L.) Kobuski.

JATIPHLA

Myristica fragrans Houtt. — Myristicaceae

JIṄGIṆĪ

Lannea coromandelica (Houtt.) Merrill. — Anacardiaceae
=*Odina woodier* Roxle.
L. grandls (Dennst.) Engl.

JĪMŪTAKA

Luffa echinata Roxb — Cucurlitaceae
=*L. echinata* var. *longistyla* Cl.

JĪRAKA

ŚVETA *Cuminum cyminium* Linn. — Apiaceae
KṚṢṆA *Carum carvi* Linn.

JIVAKA–JIVAKA—ṚṢBHAKA

Malaxis acuminata D. Don. — Orchidaceae
=*Microstylis wallichii* Lindl.
M. nusifera Ridley.

JĪVANTĪ

JĪVANTĪŚĀKA *Leptadenia reticulata* W. & A — Asclepiadaceae
L. pyrotechnica (Forsk.) Decne.
=*L. spartium* W. & A.
SVARṆAJĪVANTĪ *Flickingera macraei* (Lindl.) Bennet.
=*Dendrobium maeraei* Lindl.
F. fimbriata (Bl.) Howkes.

=*Dendrobium macraai* auct. non Lindl.
=*Desmostrichum fimbriatum* Bl.
Pholidota articulata Lindl var. *griffithü* (Hook f.) king. & Pantl.
=*P. griffithü* Hook f.
P. imbicata Lindl.

JYOTIṢMATĪ

Celastrus paniculatus Willd. Celastraceae

JŪRṆĀ

Thysanolaena agrostis Nees. Poaceae
T. procera Mez.

JAMBŪ

RĀJAJAMBŪ *Syzygium cumini* (L.) Skeels. Myrataceae
=*Eugenia jambolana* Lamk.

NADĪJAMBŪ *Syzygium operculatum* (Roxb) Niedenzu
=*E. operculatus* (Roxb) Merrill.
=*Cleistocalyx operculatus* (Roxb) Merrill.
Syzygium operculatum var. *obovatum* (Duthie) Gamble.
=*E. opercuata* var. *obovata* Duthie..
=*C. operculatus* var. *oboratus* (D.) S. & W.

JALAJAMBUKA *Syzygium heyneanum* (Duthie) wall.
=*E. heyaneana* Duthie.

JAPĀ

Hibiscus rosasinensis Linn. Malvaceae
H. rosasinensis L. var. *illiflorus* Hochr.
=*H. arnotti* Griffith.
Hibiscus syriacus Linn.

JAṬĀMĀṀSĪ

Nardostachys grandiflora Dc. Valerianaceae
=*Valeriana jatamansi* auct. non Jones.

=*N. jatamansi* Dc.

JAYAPĀLA

Croton tiglium Linn. Euphorbiaceae

JALANIMBA—NĪRĀBRĀHMĪ

Bacopa monnieri (L) Pennel.
=*Gratiola monnieria* L.
=*Herpestis monnieria* (L.) Kunth.
=*Bacopa monnieria* (L.) Wettst.

JHAṆḌU

Tagetes erecta Linn. Asteraceae

JHIṄJHĪRIṬĀ

Triumfetta rhomboidea Jacq Tiliaceae
=*T. bartramia* L.
Triumfetta pentandra A. Rich.
=*T. meglecta* W. & A.

ṬANKA

Pyrus commumis Lis Rosaceae

ṬANKARĪ

Physalis minima Linn. Solanceae

DIṆḌIŚA

Citrullus vulgaris var *fistulosus* (stocks) Duthine & Fuller

TAGARA

Valeriana jatamansi Jones. Valerianaceae
=*V. wallichü* De
Valeriana jaeschkei cl. var.
kaschmiriensis (Grubov.) Nasir
=V. *officinalis* cl auct. non L.
V. officinalis Linn.
V. leschenaultü Dc.
V. hardwickü wall.

V. himalayana Grul.
=*V. dioica* anct. non. L.

TAṆḌULĪYA (KA)

Amaranthus spinosus Linn. Amaranthacear
Amaranthus viridis L.
=*A. gracilis* Desf.
A. caudatus Linn.
=*A. paniculatus* Linn.

TAMĀLA (PATRA)—PATRA (KA)

Cinnamomum tamala Ness & Eberm. Lauraceae
C. zeylanica Blume.

TARUTA

=*D. glabra* auct. non Roxb.
Dioscrea belophylla Dioscoreaceae

TUVARI

Eruca vesicaria (L.) Cav, Cruciferae
=*Eruca sativa* Mill

TARKĀRī

Clerodendron phlomidis Linn. Verbenaceae

TAVAKṢĪRI
TIKHURA

Curcuma angustifolia Roxb.

TAMALAKI—BHUMYĀMALAKĪ

Phyllanthes niruri Linn. Euphorbiaceac
P. fratermus Webster.
P. niruri sensu Hook. f.
P. urinaria Linn.

TĀMBULA—NĀGAVALLĪ

Piper betle Linn. Piperaceae

TĀLA

Borassus flabelifer Linn. Palmae
Curculigo orchioides Gaertn. Amaryllidaceae

TĀLĪŚA

Taxus baccata Linn. Taxaceae
Abies webbiana Lindl.
Rhododendron campanulatum D. Don. ssp. *aeruginosum* (Hook.) Chemberian
R. anthopogon D. Don
=*R. haemonium* Balf. f. & Cooper.
R. anthopogon ssp. *hybenanthum* (Bail f.) Cu.
=*R. anthopogon* var. *hypenanthum* (Bail. f.) Culler Hara
R. lepidotum Wall ex G. Don.
=*R cremnastes* Bail f. & Farrer.

TUVARAKA

Hydnocarpus kurzü (king) warb. Flacourtiaceae
=*Taraktogenos Kurzü* (King)
=*H. heterophyllus Kurz.*
H. laurifolia (Dennst) sleumer
=*H. wightiana* Bl.

TUDA

Morus indica Griff. Moraceac
M. australis Poiret.
=*M. indica* auct. non. L.
M. acedosa Griff.

TRAPUṢA

Cucumīs sativus Linn. Cucurbitaceae
Gentiana kurroa Royle. Gentianaceae
Gentiana decumbens Linn. f.
G. olivieri Grisala.
=*G. daurica* Fisch.
G. tenella Pries.

Delphinium zalil Aitch. & Hemsl.
Thalictrum foliosum Dc.

TRIVṚT (Ā)

Operculina turpethum (Linn.) Silva Manso.
=*Ipomoea turpethum* (L.) R. Br.

TVAKA—DĀRUSITA

Cinnamomum zeylanicum Blume. Lauraceae
C. inera Reinw.
C. javanicum Blume.

TVAKPATRA

Cinnamomum cassia Blume Lauraceae
=*C. aromaticum* Ness.

TUNNAKA

Toona ciliate R. Meliaceae
=*Cedrela toona* Roxb. ex Rotte.
=*C toona* var. *līsteri C. Dc.*
Toona serrata (Royle) Roem.
=*Cedrela* serrata Royle.
=*C. toona* Hiern.

TULASĪ

Ocimum sancium Linn. Lamiaceae

SVETA KRSNATULSI

O. basilicum Linn.
O. pilosum willd.
Kalitulsi *Ocimum americanum* L.
=*O. canum* Sims.
=*O. africanum* Lour.
Ramtulsi—*O. gratissimum* Linn.
Kapurtulsi *O. kilimand scharicum* Guerke.

TURUṢKA

Liquidamber orientalis Miller Hamemelidaceae

L. styraciflua Linn.
Atingla excelsa Noronha.

TĀMRAKUTA

Nicotiana tobacum Linn. Solanaceae
N. rustica Linn.
Lobelia nicotinaefolia Roth. ex R.S.
=*L. pyramidalis* wall.
=*L. rosea* wall.
=*L. colorate* wall.
=*L.* erecta Hook. F. & Thoms.
=*L.* beddomcana wimmer.
=*L. pyramidalis* wall.
=*L. leschenaultiana* (Presl.) Skottsli.
=*L. nicotianaefolia* non R. & S.
=*L. inflata* Linn.
=*L. chinensis* Lour.
=*L. radicans* Thunb
=*L. heyneana* Roem. & Schult.
=*L. trialata* Ham. ex. D. Don.
=*L. zeylanica* auct. non L.
=*L. nicotianaefolia* var *trichandra* (w.) *cl.*
=*L. nicotianaefolia* non Roth. ex. R. & S.

TRIPUṬĪ

Lathyrus sativus Linn. f. Fabaceae

TINIṢĀ

Ougeinia dalbergioidus Benth. Fabaceae

TINTIDĪKA

Rhus parviflora Roxb. Anacardiaceae

TINDUKA--TINDUKĪ

Diospyros exculpta Buch.—Ham. Ebenaceae
=*D. tomentosa* Roxb.
D. melanoxylon Roxb.
D. malabarica (Desr.) Kostel.

=*D. embryopteris* Pers.
=*D. pergerina* sensu Gurke.

TILA

Sesamum orientale L.
=*Sesammm indicum* Linn. Pedaliaceae

TILAKA

Wendlandia heynei (R. & S.) Santapau & Merchant. Rubiaceae
=*W. exerta* (Roxb) Dc.
=*Rondeletia exserta* Roxb.

TUMBURU

Zanthoxylum armatum Dc.
=*Z. alatum* Roxb.
Zanthoxylum acanthopodium Dc.
=Z. acanthopodium var. timbor Hook. f.
Z. altidum (Roxb.) Dc.
=*Z. hamiltonianum* wall ex Hook. f.
Z. hamiltonianum var. tomentosum Hook. f.
=*Z. ovalifolium* wight.
=Z. ovalifolium var. sepearium (W.) Hook. f.
=Z. lurkillianum Babu
=Z. hamiltonianum sensu Burkill.
Z. rhetsa (Roxb.) Dc.
=*Z. budrunga* (Roxb.) Dc.
Z. limonella (Dennst) Alston.
Z. oxyphyllum Edgew.
Z. scandens Bl.
=*Z. khasianum* Hook. f.

DANTĪ

Baliospermum montanum Muell.—Arg. Euphorbiaceae
=*B* axillare Bl.

DĀḌĪMA

Punica granatum Linn. Punicaceae

DĀRUHARIDRĀ

Berberis chitria Lindley. **Berberidaceae**
=B. aristata sensu Hook. f.
Berberis aristata Dc.
Berberis asiatica Roxb. ex Dc.
B. lycium Royb.
B. vulgaris Linn.
B. kumaunensis Royb.
=B. *vulgaris* var *aetuensis* Nook. f.
B. pachyacentha Kochne. ssp.
zabeliana (Schneider) Jafri.
=*B. zabliana* Schneider.
B. insignia Hook. f.
B. petiolaris wall. ex G. Don.
B. umballata wall.
B. wallichiana Dc.

DŪRVĀ

Cynodon dactylon (Linn.) Pers. **Poaceac**

DEVADĀRU

Cedrus deodara (Roxb.) Loud. **Pinaceae**

DEVADĀLĪ

Luffa echinata Roxb. **Cucurbitaceae**
L. *echinata* var. *longistyla Cl.*

DRĀKSĀ

Vitis vinifera Linn. **Vitaceae**

DURĀLABHĀ—DHANVAYĀSA

Fagonia cretica Linn. **Zygophyllaceae**
Fagonia indica var. *schweinfurthii* Hadidi
=*F. arabica* auct. non L.
=*F. cretica* auct. non L.

DUGDHIKĀ

Emphorbia thymifolia Linn. **Emphorbiaceae**
Euphorbia hirta Linn.

E. orbiculata H.B.K.
=*E. microphylla* Heyne.
E. hypercifolia Linn.
E. indica Lamk.
=*E. hypercifolia* auct. non *L.*

DVĪPĀNTARA VACĀ

Smilax china Linn. Smilacaceae
S. lanceaefolia Roxb.
Smilax glabra Roxb.
S. ovalifolia Roxb.
=*S. macrophylla.* Roxb.
=*S. zeylanica* Linn.
S. ornata Hook.
S. aspera Linn.

DROṆAPUṢPĪ

Leucas caphalotes Spreng. Lamiaceae
Leucas *indica* (L.) R. Br.
=*Leonurus indicus* L.
Leucas lavandulifolia Smith. L linifolia (Roth.) Spreng.
Leucas indica var. *decipiens* (Hook. f.) Bennet.
L. plukenetü (Roth.)
=*Phlomis plukenetü* (*Roth.*) Sprang.
=*Phlomis plukenetü* Roth.
=*L. aspera* Link.
L. martinicansis R. Br.
L. stelligera wall.
L. urticaefolia R. Br.
L. zeylanica R. Br.

DAMANAKA

A. indica willd.
=*A. vulgaris* auct. non L.
Artemisia vulgaris Linn. Asteraceae

DHATTURA

Datura metel Linn. Solanaceae
=*D. fastuosa* L.

=*D. fastuosa* var. alba (Nees) Cl.
=*D. innoxia* Mill.
=*D. metel* sensu cl.
=*D. stramonium* L.
=*D. stramonium* var. *tatula* (L.) Cl.

DHANVANA

Grewia tiliaefolia Vahl. Tiliaceae.

DHAVA

Anogeissus latifolia (Roxb. ex Dc.) wall ex Guill & Perr. Combretaceae
=*A. latifolia* var. *tomentosa* Haines.
Anogeissus acuminata (Roxb. Dc.) wall ex Guill &Perr var. *phillyrefolia* (Heurck & Muell Arg.) Kurz.
A acuminata (Roxb. ex Dc) Guill & Perr. var. *acuminata*
=*A. acuminata* var. *typica* Cl.
A. sericea var. *acuminata* King ex Duthie.
=*A rotundifolia* Blatter & Haith.
A. sericea Brandis. var. *sericea*
A. latifolia var. *parviflora* Cl.
A. pendula Edgew.

DHĀTAKĪ

Woodfordia fruticosa Kurz. Lythraceae
=*W floribunda* Salisb.

DHĀNYAKA

Coriandrum sativum Linn. Apiaceac

DHĀMĀRGAVA

Luffa cylindrica (Linn.) M. Rocm. Cucurbitaceae

DHANYARUKA

Cleviceps purpurea Tul. Hypocreales
=*Sacale cornutum* Nees.

NALA

Phragmites maxima Blatter & Mc Cann. Poaceae
P. kirka Trin.
Lobelia leschenaultiana (Presl.) Skoltsle.
=*L. excelsa* Lesch.
=*L. nicotionaefolia* non R. & S.

NAGAKESARA

Mesua ferrea Linn. Guttiferae

NĀGABALĀ

Sida cordata (Burm. f.) Borssum. Malvaceae
= *Sida veronicaefolia* Lam.
=*S. humilis* cau.
Sida alba L.
=*S. spinosa* L.
Grewia tenax (Forsk.) Pioari.
=*G. populifolia* Nohl.
G. helicterifolia wall, ex G. Don.
=*G. hirsuta* var. *helicterifolia* (wall ex G. Don) Haines.

NĀRIKELA

Cocos nucifera Linn. Palmae

NIKOCAKA

Aesculus indica Coleler. Hippocastanaceae
Aesculus hippocastanum Linn.

NICULA–HIJJALA

Barringtonia acutangula (Linn.) Gaertn. Lecythidaceae

NIMBA

Azadirachta india A. Juss. Mediaceae
=*Melia azadirachta* L.

NIRGUṆḌĪ

Vitex negundo Linn. Vitaceae
V. trifolia Linn.

NĪLĪ – NĪLIKĀ

Indigofera tinctoria Linn

VANYANILA

Indigofera oblongifolia Forsk.
I. paucifolia Delile.

NĀGADAMANĪ – ĪSWARĪ

Aristolochia indica Lim. Aristolochiaceae

NĀKULĪ—(ĪŚAVRĪ)

Aristolochia indica Linn. Aristolochiaceae
Aristolochia bracteolata Lamk.
=*A. bracteata* Retz.
A. tagala Cham.
=*A. roxburghiana* Lank.

NAGADMANI (NA)

Crinum asiaticum Linn. Amaryllidaceae
C. defixum Ker.—Gawl.
Pupalia lappacea (L.) Juss.
= *Achyranthes lappacea* L.
=*P. atropurpurea* (Lamk.) Moq.
P. lappacea var. *orbiculata* (Heyne ex Wall) Townsend.
A. orbiculata Heyne ex Wall.
=*P. orbiculata* (Heyne ex wall.) W.
Sensevieria/roxburghiana Schult.

NANDĪVRKṢA

Ficus microcarpa L. Moraceae
=*F. retusa* sensu Hook. f.
F. retusa Linn. f.

NIMBUKA

Citrus aurantifolia (christin.) Swingle. Rutaceae
=*Citrus medica* var. *acida*
C. medica Linn. C.
=*C. medica* var *medica proper*

C. aurantium Linn.
var. *lergadia* Hook. f.
C. limon (L.) Burm. f.
C medica ssp. limon L.
=*C. medica* var. *limonum* Hook. f.
C. grandis osbeck.
C. dacumna L.

NARTAKA - RAGI (MAṬṬAKA)

Eleusine corcana Gaertn. Poaceae

Dactyloctenium aegyptium Beauw.
=*E. hagyptiacea* Desf.
E. indica Gaertn.

NIRGUṆḌĪ-VANDĀKA

Alectra parasitica A. Rich. var. *chitrakutensis* Scophulariaceae

(Host plant *Vitex negundo* Linn.)
Alectra sessiliflora (Vahl.) O. Kuntze.
=*Melasme arvense* (Benth.) Hand-Mazz.
=*A. indica Benth.*

PĪTAKŪṢMĀṆḌA

Cucurbita maxima Duch. Cucurbitaceae
=*C. pepo* Linn

PAṬOLA

Trichosanthes dioica Roxb. Cucurlitaceae
Vanpatola Trichosanthes cucumerina L.
=*T. brevibracteata* Kundu.

PATTANGA

Caesalpinia sappan Linn. Caesalpiniaceae

PADMAKA

Prunus cerasoides D. Don. Rosaceae
=*P. puddum* (Wall.) Roxb. ex Brandis

PANASA

Artocaspus heterophyllas Lamk. Moraceae

=*A. integrifolia* auct. mult non L.

PARŪṢAKA

Grewia asiatica Linn. Tiliaceae

=*G. suleinaequalis* Dc.

PARPATAKA

Fumariaindica Pugsley. Fumariaceae

=*F. parviflora* auct. non Lank.

F. officinalis Linn.

F. parviflora Lamk.

Fumaria vaillantii Loisel.

=*F. parviflora* ssp. *vaillanti* (Loisel) Hook. f.

Hedyotis biflora (L.) Lamk.

=*Oldenlandia biflora* L.

=*O. paniculata* L.

=*Gonotheca biflora* (L) Masam.

=*Thecogonum biflorum* (L) Babu.

H. corymbosa (L.) Lamk.

=*O. corymbosa* L.

=*O. burmanniana* R. Br.

H. diffusa (willd.)

=*O. aiffusa* (willd.)

H. herlacea L.

=*O. hcrbacea* (L.) Roxbe.

Mollugo cerviana Ser.

Glinus lotoides L.

=*M. hirta* Thunb.

M. pentaphylla L.

=*M. strieta* L.

M. nudicaulis Lam.

Polycarpaea aurea W. & A.

=*P. corymbosa* var. *aurea* (W. & A.) W.

P. corymbosa Lam.

Glossocardia bosvallia Dc.

Justicia procumbens Linn.

J. *japonica* Thunb.

=*J. simplex* D. Don.

=*J. procumbens* var. *simplex* (D. Don) Yamazaki.

PALĀŚA

Butea monosperma (Lam.) Kuntze. Fabaceae
B. monosperma (Lamk.) Taub. var. *monosperme* Maheshwari.
=*B frondosa* Koen. ex Roxb.

PĪTAPĀLAŚA

Butea monospesma var *lutea* (Witt). Maheshwari
=*B. frondosa* Koen. ex Roxb. var. *lutea* Witt.
=*B. lutea* Sagreiya.

PALĀNḌU

Allium cepa Linn. Liliaceae

PATRASNUHI

Opuntia stricta (Haw) Haw. var. *dillenü* (Ker-Gouol.) Benson. Cactaceae
=*O. dillenü* (Ker-Gawler) Haw.

PĀṬALĀ

Stereospermum suaveolevs Dc. Bignoniaceae
StereosPermum Chelonoides (Linn. f.) Dc.
=*Bignonia suavelens* Roxb.
=S. *suaveolens* (Roxb.) Dc.
Stereospermum colas var. *angustifolium* (Haines) Bennet & Raizada
=*S. tetragonum* Dc.
=*S. personatum* var. *angustifolium* (Haines) Santisuk.

PĀṬHĀ

Cissampelos pariera Linn. Menispermaceae
C. pareira L. var *hirsuta* (Buch-Ham (Dc.) Forman.
C. pareira L.
=*C. nepalensis* Rhods.
Cyclea peltata (Lamk.) Hook. f. & Thoms.
=Syn. *C.* barbata Miers.
=*C. burmenü* Hook. f. & Thoms.
Stephania japonica (Thunb.) Miers.
=*S. hernandifolia* (willd.) Walp.

PĀRIJĀTA—ŚEPHĀLĪKĀ

Nyctanthes arbortristis Linn. Obaceae

PĀRIBHADRA

Erythrina variegata L: Fabaceae

=*E. indica* Lank.

=E. variegata L. var. orientalis (L.) Merrill.

=E. variegata L. var. *orientalis* fa. *orientalis* (L) Mahes.

Erythrina suberosa Roxb

=*E. glabrascens* (Prain) Parker.

E. stricta Roxb.

PALAKYĀ

Spinacia oleracea Linn. Chenopodiaceae

PĀṢĀṆABHEDA

Bergenia ligulata (wall.) Engl. Saxifragaceac

=Saxifraga *lingulata* wall.

Bergenia ciliata (How.) Sternlo.

=*Saxifraga ligulata* Var. *ciliata* (Royle.) Hook. f. & Thoms

B. ciliata fa. *ligulata* yeo

=*B. ligulata* sensu Bailey.

B. starcheyi (Hook. f. & Thoms.) Engl.

=*S. stracheyi* (Hook. f. & Thoms.

Aerva lanata Juss.

PARṆABĪJA

Kalanchoe pinnata (Lank.) Pers.

=*Bryophyllum calycinum* Salisb.

=*B. pinnatum* (Lamk.) Oken.

K. lacinata (L.) Dc.

=*K. schweinfurthü* Penzig.

K. integra (Medik.) Kuntze.

=*K. spathulata* Dc.

Coleus ambonicus Lour.

=*C. aromaticus* Benth.

Homonoia riparia Lour.

Rotula aquatica Lour.

=*Rhabdia lycioides* sensu cl.

Nomocharis brachiata bt.

Kiekxia ramosissima (wall) Janchen.

=*Linaria ramosissima* wall.

PIPPALI

Piper longum Linn. Piperaceae

PĪLU

KṢUDRA *Salvadora oboides* Dcne. Salvadoraceae

BṚHATA *S. persica* Linn.

=*S. indica* Wight.

PILUPARṆI

Maerua oblongifolia var. *scabra* (Hook. f. & Thoms.) Nair. Capparidaceae

=*M. arnenaria* var. *glabra* Hook. f & Thoms.

M. oblongifolia var. *scabra* (Hook. f. Thoms.) Nair

=*M. arnenaria* var. *scabra* Hook. f. & Thoms.

M. oblongifolia (Forsk.) A. Rich.

=*M. arneraria* (Dc) Hook. f. & Th.

Clematis gouriana Roxb.

Salvadora oleboides Dcne.

PUTRAJIVAKA

Putranjiva roxburghii wall. Euphorbiaceae

PUNARNAVĀ

Bocrhavia diffusa L. Nyctaginaceae

RAKTA

=*B. repens* L.

=*B. repens* var. *diffusa* (L.) Hook. f.

=*B. repens* var. *procumbens* (Roxb.) Hook. *f*,

B. rependa willd.

ŚVETA

Trianthema decandra Linn. Ficoidaceae

T. portulacastrum L.

=*T. monogyna* L.

T. triquetra Rottl. ex willd.

=*T. crystallina* auct. non vahl.

T. pentendra Linn.

PUNNĀGA

Calophyllum inophyllum Linn. Guttiferea

SURAPUNNĀGA

Orchrocarpus longifolius Benth.

PIŚĀCAKĀRPĀSA

Abroma agusta Linn. Sterculiaceae

PUṢKRMULA

Inula racemosa Hook f. Asteraceae

I. royleana Dc.

PUGA

Areca Catechu Linn. Palmae

PUTIKARANJA—KARANJA

GHṚTAKARANJA *Pongamia pinnata* (Linn.) Merr. Fabaceae

Derris indica (Lamk.) Bennet.

=*Pongamia glabra* Vent.

=*P. pinnata* (L.) Pierre.

non *Derris pinnata* Lour.

KAṆṬAKARAÑJA

Caesalpinia bonduc (L.) Roxb. Caesalpiniaceae

=*C. bonducella* (L.) Fleming.

=*C. major* (Medik.) Dandy & Exell.

=*C. bonduc* sensu Roxb.

C. crista L.

=*C. bonducella* (Linn.) Fleming.

PṚŚNIPARṆĪ

Uraria picta Desv. Fabaceae

U. lagopoides Dc

PRASĀRIṆĪ

GANDHAPRASĀRIṆĪ *Paederia foetida* Linn. Rubiaceae

Leptadenia pyrotechnica (Forsk.) Decne.

=*L spartium* W. & A.

Paederia scandens (Lour.) Merrill.

P. tomentosa Bl.

PĀRASIKATAGARA

Asarum europacum Linn. Aristolochiaceae

PRIYAṄGU

Callicarpa macrophylla Vahl. Verbenaceae
C. nudiflora Hook. & A.
=*C. macrophylla* var. *sinensis Cl.*
Prunus mahalele Linn.
Aglaia elaeagnoidea (Juss.) Benth.
=*A. roxburghiana* (W. & A.) *Miq.*
A. elaeagnoides var. *Courtallensis* (Gamble) Nair.
=*A. roxburghiana* var. *coutrallensis* Gamble.

PRIYĀLA

Buchanania lanzan Spr. Anacardiaceae
=*B, latifolia* Roxle.

PLAKṢA

Ficus lacor Buch.-Ham. Moraceae
Ficus arottiana Miq.
F. rumphü Blume.
F. tablotti G. King.

PĀRASĪKAYANĀNI

Hyoscymus niger Linn. Solanaceae
Hyoscymus muticus Linn
H. pusillus Linn.
H. reticulatus Linn.

PĀRASIKAVACĀ

Iris germanica Linn. Iridaceae
Paris polyphylla Sm.
Iris Kumaonensis wall.

PARPAṬĪ

Pogostemon heyneanthus Benth. Laminaceae
=*P. patchouli* sensu Hook. f.

PĀTĀLGARUḌĪ—CHHILAHIṆṬA

Cocculus hirsutus (Linn) Diels. Menispermaceae
=*Mennispermum hirsutum* L.
=*C. villosus* Dc.
Cocculus orbiculatus (L.) De.
=*Mennispermum orbiculatum* L.
=*C. mollis* wall. ex Hook. f. & Thoms.
=*C. trilobus* (Thunb.) Dc.
=*C. sarmentosus* (Lour.) Dc.
C. *pendulius* (J.R. & G. Forst.) Diels.
=*Epibatrium pendulum* J. R. & G. Forst.
=*C. leaeba* (*Delile*) Dc.

PĀRIṢA

Thespesia populneoids (Roxb.) Kostal. Malvaceae
=*Hibiseus populneoides* Roxb.
=*T. populnea* var. *populaenoides* (Roxb.) Pierre.
=*T. populnea* sensu Masters.
T. populnea Soland ex Correa.
Thespesia populnea (L.) Corr.

PRĀCINĀMALAKA

Flacourtia jangomas (Lour.) Raeusch. Flacourtiaceae
=*Stigmarota* jangomas Lour.
=*F. catophracta* Roxb. ex willd.

PŪTIHĀ

Mentha longifolia (L.) Huds. Lamiaceae
=*M. longifolia* var. *incana* (Willd.) Raizada & Saxena
=*M. longifolia* var. royleana (Benth.) R. & S.
=*M. spicata* L. var. *lengifolia* L.
=*M. sylvestris* L. var. incana (Willd.) Hook. f.
M. arvensis Linn.
M. piperita Linn.
M. sativa Linn.
M. viridis Linn.

PHAṄJĪ

Rivea ornata Chois Convolvulaceae

PHALGU

Ficus carica Linn. Moraceae
F. hispida Linn. f.

BAKULA

Mimusops elengi Linn. Sapotaceae

BANDHUKA–BANDHUJIVA

Pentapetes phoenicea Linn. Sterculiaceee

BAHIRŚIKHĀ

Elephantopes scaber Linn. Asteraceae

BALĀ

Sida cordifolia Linn. Malvaceae
Sida acuta Burm. f.
=*S. carpinifolia* sensu Masters.
S. alba Linn.
=*S. spinosa* L.
S. rhombifolia ssp. *retusa* (L.) Borssum.
=*S. rhombifolia* var *retusa* (L.) Masters.
S. rhombifolia L. ssp. rhombifolia Masters.
=*S. orientalis* Cav.

BĀKUCĪ

Valeriana wallichii Dc. Valerianaceae
Coleus vettiveroides K. C. Jacob.

BIBHITAKA

Terminalia bellirica (Gaertn) Roxb Combretaceae
=*T. belerica* Roxb.

BIMBI

Coccinia grandis (L.) Voigt. Cucurbitaceae
=*C. cordifolia* Cogn.
=*C. indica* W. & A.

BILVA

Aegle marmelos Corr.

BĪJAPŪRA (KA)

Citrus medica Linn. Rutaceae

BRĀHMI

Bacopa monnieri (Linn.) Pennel. Scrophulariaceae
=*Herpestis monnieria* (Linn.) H. B. & K.
=*B. monnieria* (L.) mettst.

MAṆḌUKAPARṆĪ

Centella asiatica (Linn.) Urban. Apiaceae
=*Hydrocotyle asiatica* L.
H. javanica Thunli.
=*H. himalaica* Muker.
H. sibthorplaides Lank.
=*H. rotundifolia* Roxb.

BABBULA

Acacia nilotica ssp. *indica* (Benth.) Brennan.
Mimosaceae
=*A. arabica* (Lmk) willd.
=*A. nilotica* var. *indica* (Benth.) Hill.
=*A. arabica* sensu Baker.
Acacia nilotica ssp. *cupressiformis* (Stewart.) Ali & Faruqi.
=*A. nilotica* ssp. *indica* var. *cupressiformis* (Stewart) V. & K.
A. nilotica (L.) Delia ssp. *astringens* (Schum & Thoms) Roberty.
=*A. nilotica* var. *adansonü* (Gill & Perr.) Brenau.
A. ferruginea Dc.
A. jacquemontü Benth.
A. leucophloca Willd.

BARBARĪ

Ocimum basilicum Linn. Laminaceae

BRAHMADAṆḌĪ

Tricholepis glaberima Dc. Asteraceae
Tricholepis angustifolia Dc.

BARJARI

Pennisetum typhoides Rich. Poaceae

BHALLĀTAKA

Semecarpus anācardium Linn. Anacardiaceae

BHAVYA

Dillenia indica Linn Dilleniaceae

BHARGI (BHĀRAṄGĪ)

Clerodendrum serratum (Linn.) Moon. Verbenaceae
C. indicum (L.) O. Kuntze.
C. siphonanthes R. Br.

[*Clerodendrum*=*Clerdandron*]

Picrasma quasssioidus Benn.

P. javanica Bl.

=*P. javanica* var. *nepalensis* (Benn.) Badlwar.

Gardenia latifolia Ait.

G. turgida Roxb.

G. resinifera Roth.

=*G. lucida* Roxb.
Clerodendrum viscosum Vent.
=*C. infortunatum* auct. non L.
=*Clerodendrum viscosum* Vent.
C. infortunatum Linn.
Pygmacopremna herbacea (Roxb.) Moldenke.
=*Premna herbacea* Roxb.

BHILLOTA (KA)

Euonymus tingens wall. Celastraceae
E. pendulous wall.

BHUTAKEŚĪ MURĀMĀṀSĪ

Selinium condollü Dc. Apiaceae
=*S. filicifolium* (Edge.) Nasir.
=*S.* wallichianum (Dc.) Raizada & Saxena.

=*S. tenuifolium* Wall excl.
S. viginatum C.B. Clarke.

BHURJA

Betula utilis D. Don. — Betulaceae
=*B. bhojpattra* wall.
Betula alnoides Buch-Ham.
=*B. acuminata* wall.

BHṚNGARĀJA

Eclipta prostrata (L.) L. — Asteraceae
=*E. alba* (L.) Hassk.
Wedelia calendulacea Less.
W. wallichü Less.

BHAṆḌĪRA

Clerodendrum viscosum Vent. — Verbenaceae
=*Clerodendrum infortunatum* Linn.

BHUDAṆḌĪ

Amberoba ramosa (Roxb.) Jafer. — Asteraceae
=*Volutarella rivaricata* (Dc.) Benth.
=*V. ramosa* (Rox.) Sant.
=*Oligochaeta ramosa* (Roxb.) Wagenitz.

MAKUṢṬHA(KA)

Phaseolus aconitifolius Jacq. — Fabaceae

MANJIṢṬHĀ

Rubia cordifolia Linn. — Rubiaceae

MAṆḌUKI (MAṆḌUKAPARṆI)

Centella asiatica (Linn.) Urlan. — Apiaceae
=*Hydrocotyle asiatica* Linn.

MADANA (PHALA)

Catunaregam spinosa (Thunb.) Tiruv. — Rubiaceae
=*Randia dumetorum* (Retz.) Poir

=*R. brandisü* Gamble.
=*Xeromphis spinosa* (Thumb.) Keay.
C. nutans (Dc) Tiruveng.
=*R. nutans* Roxb.
=*R. dumetorum* auct. non Lamk.

PIṆḌITAKA

Catunaregam uliginosa (Retz.) Sivarajan.
=*Randia uliginosa* (Retz.) Poir.
=*Xeromoph's uliginosa* (Retz.) Mahes.

MADAYANTIKĀ

Lawsonia inermis Linn. Lythraceae
=*L. alba* Lamk.

MADHŪYAṢṬI

Glycyrrhiza glabra Linn. Fabaceae

MADHUKA

Madhuca indica J.F. Gmel. Sapotaceae
=*Madhuca latifolia* (Roxb.) Maclride.
Madhuca longifolia (Koenig.) McBride.
=*Bassia longifolia* Koenig.
=*M. latifolia* Roxb.

MADHUKAPUṢPI

Diploknema butyracea (Roxb.) Lam. Sapotaceae
=*Bassia butyracea* Roxb.
=*Madhuca butyracea* (Roxb.) Macbride.
=*Aesandra butyracea* (Roxli.) Bachni.
Diplokema butyraceoides (Scott.) Lam.
=*Bassia butyraceoides* Scott.

MARICA

Piper nigrum. Linn. Piperaceae

MARUBAKA

Origanum vulgare L Lamiaceae
Majorana hortensis Moench.
=*Origanum majorana* Linn.

MALLIKĀ

Jasminum sambac Ait. Oleaceae

MASURA

Lens culinaris Medic. Fabaceae

=*Erum lens* L.

=*L. culinaris* ssp. *nigricans* (Bich.) Thellung.

MAHĀNIMBA

Melia azedarach Linn. Meliaceae

Ailanthus excelsa Roxle.

MACIKA

Tamarix aphylla (L.) Karst. Tamaricaceae

=*T. articulata* Vahl.

Tamarix gallica Linn.

=*T. troupii* Hole.

T. troupii Hole.

=*T. gallica* auct. non L.

MĀNAKA

Alocacia indica Schott. Araceae

Alocacia macrorhiza (L.) G. Don.

=*Arum macrorhizum* L.

=*Alocacia indica* (Lour.) Spach.

MARṢA MĀRIṢA

Amaranthus blitum Linn. Amaranthaceae

=*A. lividus* L.

Amaranthus lelitum var *oleracea* Duthine.

A. tricolor L.

=*A. gangeticus* L.

=*A.* mangostanus Linn.

MĀLATĪ

Aganosma dichotoma (Roth.) K. Schum. Apocynaceae

MĀṢA

Vidana mungo (L.) Hepper. Faleaceae
=*Phaseolus mungo* Linn.

MUDGA

Vigna radiata (L.) Wilez Faleaceae
=*Phaseolus radiatus* Linn.
=*P. aureus* Roxb.
=*P. mungo* var. *radiatus* (L.) Baker.

MĀṢAPARṆI

Teramnus labialis Spreng. Fabaceae
T. mollis Benth.
=*T. labialis* var *mollis* Baker.

MUCAKUNDA

Pterospermum acerifolium Willd. Sterculiaceae
Pterospermum canescens Roxb.
=*P.* suberifolium (L.) Lamk.
P. xylocarpum (Gaertn.) Sant. & Wagh.
=*P. heyneanum* Wall ex W. & A.

MUNJĀTAKA

SĀLAM *Orchis latifolia* Linn.
O. masculata Limm.
Dactylorhiza hatagirea (D. Don.) Soo.
=*O.latifolia* auct. non L. Orchidaceae

MUDGAPARṆĪ

Pueraria lobata (Willd.) Ohwi. Fabaceae
=*Phaseolus trilobus* (L.) Ait.
=*Pueraria trilolea* (L.) Makino.
=*Dolīchos lobatus* Willd.
Vigna trilolata (L.) Verdẹ.
=*Phaseolus trilolus* sensu Aiton.

MUŚALĪ

Chlorophytum tuberosum Baker. Liliaceae
C. arundinaceum Baker.

MUṢKAKA—MOKṢAKA

SVETA *Schrebera swietenioides* (Roxb.) Oleaceae

KṚSNA *Cassine glauca* (Rottle) O. Ktze.

=*Elaeodendron glaucum* (Rottle.) Pers.

C. paniculata (W. & A.) L. Callen.

=*E. glaucum* auct. non (Rottle.) Pers.

MUSTAKA

Cyperus rotundus Linn. Cyperaceae

Cyperus rotundus L. ssp. *retzil* Kuk

=*C. retzü* Neess.

=*C. tuberosus* sensu cl.

C. scariosus R. Br.

MURVĀ

Marsdenia tenacissima W. & A. Asclepiadaceae

Chonemorpha fragrans (Moon.) Alston.

=*C. macrophylla* (Roxb) G. Don.

Bauhinia vahlü w. & A.

Sansevieria roxburghiana Schult.

Maerua arenaria Hook. f. & thowo

Crinum defixum Kar.-*Gawl.*

MULAKA

Raphanus sativus Linn. Cruciferae

MUṢIKAPA (KA) RNI

Merremia gangetica (L.) cufo

=*Ipomoca reniformis* (Roxb.) Chois.

=*M. emarginata* (Burm. f,) Hallis.

Hiercium pilosella Linn.

H. virosum Pall. Convolvulaceae

MEDĀ-MAHĀMEDĀ

Polygonatum verticillatum Allioni. Liliaceae

P. multiflorum Allioni.

P. cirrhifolium Royle.

MEṢAṢRNGĪ

Dolic handrone folcata Seem. Bignoniaceae
Gymnema sylvestra R. Br.

MATSYĀKṢĪ

Alternanthera sessilis (Linn.) R. Br. ex Dc. Amananthaceae
=*Gomphrena sessilis* L.
=*A. modiflora* R. Br.

MUNḌI—ŚRĀVAṆĪ

Sphaeranthus indicus L. Asteraceae
S. senegalensis Dc.
=*S. indicus* auct. non L.

MĀṀSAROHIṆĪ Meliaceae

Soymida febrifusa A. Juss.

MADHUNĀŚINĪ

Gymnema sylvestre (Ritz.) Schult. Asclepiadaceae

MUNJĀTAKA

Orchis latifolia Hook. f. Orchidaceae
Daetolyrhiza hatagirea (D. Don.) Soo.
=*Orchis hatagirea* (D. D.)
=*O. latifolia* auct. non L., Hook. f.
Eulophia compestris Wall.
Salampanja—*Eulophia dabia* (D. Don.) Hochr.
=*Eulophia compestris* Lindl.
=*Bletia debia* D. Don.

METHIKĀ

Trigonella fornumgraceum Linn. Fabeaceae
Vanamethi—*Meliolotus indica* (L.) All.
=*Trifolium indica* L.
=*M. parviflora* Desf.
M. alba Desr.
M. officinalis Lam.

MAHĀBHARI VACĀ

KULINJANA *Alpinia galanga* Willd. Zingiberaceae
A. allughas Rose.
A. khulanjana M. Sheriff.
A. officinarum Hance.
A. calcarata Rosc.
Zingiber zerumbet Rosc ex Smith.
Vernonia cinerea Less. Arteraceae

MĀRKAṆḌIKĀ

Cassia senna L. Caesalpiniaceae
=*C. angustifolia* Vahl.

MĀYĀPHALA

Quercus infectoria Oliver.

MAYŪRSIKHĀ

Actinopteris dichotoma Bedd. Polypodiaceae
Adiantum caudatum Linn.
Celosia argentea L. fa. cristata (L.) Schinz.
=*C. cristata* L.
=*C. argentea* var. *cristata* (L.) Kuntze.
C. argentea L.
Elephantopus scaber Linn.

MAKHĀNNA

Euryale ferox Salislb. Nymphaeaceae

MADHUKARKAṬĪ

Citrus maxima (Burm.) Merr. Rutaceae
=*C. decandra* Linn.
Citrus paradisi Mact.

MIṢṬANIMBŪKA

Citrus limetta Risso. Rutaceae
=*C. medica* L. var. liemetta (Risso.) Hook. f.
=*C. limettoides* Tanaka.
Citrus sinensis (L.) Osbeck.

=*C. aurantium* var. *sinensis* L.
=*C. aurantinum* var. *aurantium* proper Hook f.

MĀMAJJAKA

Enicostema littorale Blume. Gentianaceae
=*E. veritcillatum* (L.) Engl.

MUKTAVARCĀ

Acalypha indica Linn. Euphorliaceae
Acalypha racemosa Wall ex Benth.
=*A. poniculata* lliq

YAVA

Hordeum vulgare Linn. Poaceae
ATIYAVA *Avena sativa* Linn.

YAVATIKTĀ

KĀLAMEGIHA *Andrographis paniculata* Nees. Acanthaceae
Indoneesiella echioides (L.) Streem.
=*Andrographis echioides* (L.) Nees.
=*Neesiella echioides* (L.) Sreem.

YAVĀNĪ

Trachyspermum ammi (L.) Sprague. Apiaceae
=*Carum copticum* (L.) Cl.
=*Ptychotis ajowan* Dc.

YAVĀSA (KA)

Alhagi pseudalhagi (Biele.) Desv. Faleaceae
=*A. Camelorum* Fisch.

YĀTUKA

Pergularia daemia (Forsk.) Chiov. Asclepiadaceae
=*Daemia extensa* R. Br.
=*P. extensa* N.E. Br.

THIKĀ

Jasminum auriculatum Vahl. Oleaceae

YĀVANALA

Sorghum vulgare (Linn.) Pers. Apiaceae
Sorghum vulgare (Linn.) Pers var.
saccharatum Boerl.
=*S. saccharelum* Pers.

RĀJAMĀṢA

Vigna cylindrica Skeels.

RASONA–LAUNA

Allium cepa Linn. Liliaceae

RĀJĀDANA

Manilkara hexandra
=*Mimusops hexandra* Roxb. Sapotaceae
Manilkara kauki Dule.
=*Mimusops kauki* Linn.
Manilkara zapota (L.) van. Royen.
=*M. achras* (Mill.) Fosberg.
=*Nispera achras* (Mill) Fosberg.

RĀJIKA

Brassica juncea (Linn.) Czern & Coss. Brassicaceae
B. nigra (Linn.) Koch.
Brassica juncea (L.) ezern. ssp. *juncea* var.
cuneifolia (Roxb.) Kitamura.
B. juncea ssp. *juncea* var. *integrifolia* (West.) Sinskaja
=*B. juncea* sensu Hook. f. & Anderson.

RĀSNĀ

Pluchea lanceolata Oliver & Hiern Asteraceae
Pluchea pinnatifida Hook. f.
P. indica Less.
Vanda tesselata (Roxb) G. Don.
=*Vanda roxgburghii* R. Br.
Alpinia galanga Willd.
Inula racemosa Hook. f.

Acampe papilosa (Lindl.) Lindl.
=*Saccolobium papillosum* Lindl.
A. praemorea Blatter & McLann
=*S. wightianum* (Lindl.) Hook. f.
Vanda fertacea (Lindl) Reichb f.
=*V. parviflora* Lindl.

RUDRĀKṢA

Elaeocarpus sphaericus (Gaertn.) K. Schum. Elaeocarpaceae
=*Ganitrus sphaericum* Gaertn.
=*E. ganitrus* Roxb.

Elaeocarpus petiolatus (Jack.) Wall.
=*Monocera petiolata* Jacq.
=*E. integer* Masters.
E. glaudulosus Wall ex Merrill.
=*E. oblongus* sensu W.
E. recurvatus Corner.
=*E. ferrugincus* (W.) Beddome.
E. serratus L.
=*E. Cuneatus* W.
E. tectorius (Lour.) Poir.
=*E. robustus* Roxb.
=*E. leptostachya* Wall ex c. Muell.
E. venustus Bedd.
=*E. monocera* auct. non cav.

RAJANIGANDHĀ

ŚABBU *Polianthes tuberosa* Linn. Amaryllidaceae

ROHITAKA

Tecomella undulata (G. Don.) Seem Bignoniaceae
=*Bignonia undulata* Sm.
=*Tecoma undulata* G. Don.

Aphanamixis polystachya (wall).
=Aglaia polystachya wall.
=Amora rohituka (Roxb.) W. & A.

ROHIṢA

Cymbopogon martinū (Roxb.) Wats. Poaceae
C. jawarancusa (Jones.) Schult. ssp. *olivieri* (Boiss.) Soenarko.
=*C. schoenanthus* auct. non (L.) Spreng.
=*C. olivieri* (Boiss.) Bor.
=*C. ladakhensis* Gupta.
C. schoenanthus (Linn.) spreng.
C. caesius Hook.
C. pospischillü (K. Schum.) Hubbard.
=*Andropogon nardus* L. var. *stracheyi* Hook. f.
=*C. stracheyi* (Hook. f.) Raizada
C. nardus (Linn.) Rendle.
=*A. nardus* Linn.
C. flexuosus (Steud.) wats.
=*Andropogon nardus* var. *fleauosus* Hack.

RUDANTĪ–RUDRAVANTĪ

Capporis moonü wight. Capparidaceae
Cressa cretica Linn.
Astragallus cendolleanus Royle ex Benth.
=*A. anomalus* Bunge.
Astragalus chlorostachya Lindl.
=Aedysarum strobiliferum Baker
A. leucocephalus Grah.
A. multicepa Benth.
=*A. leptocentrus* Bunge.
=*A.* jacquemonti Bunge.
A. strolbiliferus Royle.

LAKUCA

Artocarpus lakoocha Roxb. Moraceae

LATĀKASTŪRIKĀ

Hibiscus ablemoschus Linn. Malvaceae
Ablemoschus moschatus ssp. *moschatus* var. *beautifolius* (Masters) Hochr.
=*H. ablemoschus* var. *beautifolius* Masters.

LAVAṄGA

Syzygium aromaticum (Linn.) Merr. & L.M. Perry
Myrtaceae

=*Eugenia caryophylla* Thunb.

LAVALĪ

Phyllanthus acidus (L.) Skeels.
=*Cicca acida* (Linn.) Merrill. Euphorbiaceae
=*Phyllanthus distichus* Muel. Arg.
=*C. disticha* L.

LĀṄGALĪ

Gloriosa superba Linn. Liliaceae

LOṆĪ

Portulaca quadrifolia Linn Portulacaceae

P. olaracea Linn.

LODHRA

Symplocos racemosa Roxb. Symplocaceae

ŚĀVARLODHRA

=*S. racemosa* var. *khasiana* Cl.

PAṬṬIKĀ LODHRA

S. paniculata (Thunb.) Miq.

=*S. crataegioides* Buch.-Ham.

LAKṢMAṆĀ

ŚVETA KANṬAKĀRĪ

Solanum surattense Burm. f. Solanaceae

(White-flowered variety)

Remusatia vivipara Schutt.
Lasia spinosa (L.) Thw.
=*Dracontium spinosum* L.
Lasio heterophylla (Roxb) Schott.
Ipomea separia Role.

Ipomoea sepiaria Roxb.
=*I. maxima* auct. non (L. f.) Sweet.
Mandaragora officinarum Bertol.
M. autumnalis Spreng.
Panax leurkillianus Bennet & Viswas.
=*Aralia quinquefolia* var *angustifolium.*
Burkill non *Panax angustifolia* F. Muell.
=*P.* pseudoginseng var. angustifolius (Burkill) Li.
=*P. japonicus* var. *angustifolius* (Burkill) Cheng & Chau.
P. pseudoginseng wall.
=*A. pseudoginseng* sensu cl.
non *P. pseudoginsing* wall.
P. bipinnatifidus Seem.
=*P.* pseudoginseng var. *leipinnatifida* (Seem) Li.

LAJJĀLU

Mimosa pudica Linn. Mimosaceae
Biophytum sensitivum (Linn.) Dc.

LATĀPALĀSA

Butea parviflora (Roxb.) O. Kuntze. Fabeaceae
=*Spatholobus roxburghii* Benth.
=*S. parviflorus* (Roxb.) O. Kuntze.
B. superba Roxb.

LANKĀ—KAṬUVIRĀ

Capsicuum annum L. var. *annuum* Dale. Solanaceae
=*C. frutescens* sensu cl.
C. annuum L. var. *cerasiforme* (Miller.) Irish.
=*C. cerasiforme* Miller.
C. grossus willd. var. *cerasiforme* (Miller) Cl.
C. annmum L. var. *glabriusculum* (Dunal.) Halser.
=*C. mimimum* Roxb.
=*C. frutescens* var. *baccatum* Irish.
C. annuum L. var. *grossus* (Willd.) Sender.
=*C. minimum* sensu Cl.

VAṀSA

Bambusa arundinacea willd. Poaceae
=*Bambosa bamlxos* Druce.

B. vulgasis Schrad.

B. glaucescens (Willd.) Siele ex Munro.

B. kurzü (Munro) Balakr.

VACĀ

Acorus calamus Linn. Araceae

Paris polyhylla Sm.

Iris germanica L. var. *nepalensis* (Lindl.)

=*I. neplensis* Wall ex Lindl. non D. Don.

I. decora wall.

=*I. nepalensis* D. Don.

I. clarkei Baker.

=*I. himalaica* Dykes.

I. kumaonensis wall.

VAṬA

Ficus lengalensis Linn Moraceae

VATSANĀBHA

Aconitum chasmanthum Stapf. ex Holmes. Ranunculaceae

A. ferox wall.

Aconitum heterophylloides (Bruhi) Stapf. var. *leucanthum* (Bruhi) Lauener.

=*A. ferox* wall. ssp. *ferox* var. *leucanthium* (Bruhi) Stapf.

A. balfourü Stapf.

A. deinorrhizum Stapf.

A. luridum Hook. f. &. T.

A. bisma (Ham.) Rapaies.

=*A. palmatum* D. Don.

A. falconeri Stapf.

Aconitum wallichianum Louener.

=*A napellus* L. var. *rigidum* Hook. f.

VANDĀKA – VṚKṢĀDANĪ

Dendrophthoe falcata (Linn. f.) Ettingshausen.

Loranthaceae

=*Loranthus falcata* Linn. f.

=*L. longiflorus* Desr.

Dendrophthoe elastica (Desr.) Danses.

=*Loranthus elasticus* Desr.

=*D. pentendra* (Linn.) Miq.

=*L. pentandrus* Linn.

Dufrenoya granulata Hook. f. & Thoms. ex A. Dc.) Stauffer.

=*Dendrophthoe granulata* (Hook. f. & Thoms. ex A Dc.) Henry & Roy.

Viscum album Linn.

V. album L. ssp. *meridianum* (Danser.) Long.

=*V. costatum* Gamble

V. monoicum Roxb.

V. orientale willd.

V. articulatum Burm.

=*V. nepalense* Spreng.

Acampe multiflora (Lindl.) Lindl.

=*Vanda multiflora* Lindl.

=*Saccolobium lnngifolium* (Lindl.) Hook. f.

A. papilosa (Lindl.) Lindl.

=*S. papillosum* Lindl.

A. praemorsa (Roxb.) Blatt. & Mc Cann.

=*S, premorsum* (Roxb.) Hook. f.

=*S. wightianum* (Linnl.)

VARUṆA

Crateva magna (Lour.) Dc, var. *magna* Capparidaceae

=*C. religiosa* var. *nurvala* (Buch.-Ham.) Hook. f. & Thoms

=*C. lophosperma* Kurz.

=*C. nurvala* Buch.-Ham.

Crateva religiosa Forst.

=*C. nurvala* sensu Kanj. & Das. (non Buch.-Ham.)

C. adansonii Dc.

=*C. odora* Buch.-Ham.

=*C.* religiosa var. *roxburghü* (R. Br.) Hook. f. & Thoms.
=*C. nurvala* sensu Blatter.
C. unilacularis Buch.-Ham
=*C. religiosa* var. *roxburghü* (R. Br.) Hook. & Thoms.
=*C. roxburghū* sensu Kavj & Das.
[*Crateva nurvala* Buch.-Ham.
=*Crataeva religiosa* Hook. f. Thoms.]

VASUKA–BUKA

Osmanthus fragrans Lour. var. *longifolius*
Obeaceae (Dc) Hara
=*Olea acuminata* var. *longifolia* Dc.
=*Notelaea posua* D. Don.
=*Osmanthus fragrans* Lour.

VĀTĀMA—MIṢṬA VĀTĀDA

Prunus anygdalus Batsch. Rosaeeae

TIKTA VATADA

P. amygdalus var. *amara*

VĀRĀHI–GṚSṬIKĀ

Dioscorea bulbifera L. Dioscoreaceae
=*D. sativa* auct non Roxb.
=*Dioscorea sativa* Thunb. non Linn.
=*D. pulchella* Roxb.
=*D.* vesicolor Buch. Ham ex wnll.
=*D. crispata* Roxb.
Tacca aspera Roxb.
T icontopetaloides (L.) Kuntze.
=*T. pinnatifida* J. R. & G.

VĀRTĀKA–VṚNTĀKA

Solanum melongena Linn. Solanaceae

VĀSĀ

Justicia adhatoda L. Acanthaceae
=*Adhatoda zeylanica* Medicus.
=*A. vasica* Nees.

VĀSTUKA

Chenopodium album Linn. Chenopodiaceae
C. murale Linn.
C. ambersioides Linn.

VIKANKATA

Flacourtia indica (Burm. f.) Merr. Flacourtiaceae
=*F. romantchi* L,
=*F. romantchi* var. sapida (Roxb.) Hook. f. & Thoms
=(also var. *latifolia*, var, *occidentalis* Hook. f. & Y Thoms)
=*F. sepiaria* Roxb.
=*F. latifolia* Cook.
=*F. occidentalis* Blatter.

VIJAYĀ—BHANGĀ

Cannabis sativa Linn. ssp. *indica* var. indica (Lamk.) Vehmer.
=*C. indica* Lamk.
=*C. sativa* sensu Hook. f.
=*C. sativa* var. *indica* Pers.
=*C. sativa* ssp. *indica* var. *kafiristanica* (Vavilov) Small & *cronquist*.

VAIJAYANTI

Canna orientalis Rose.
=*Canna indica* Linn. Cannaceae
=*C. chinensis* willd.
=C. *indica* var.
nepalensis (Bouche) Baker.
C. coccinia Mill.
=*C. indica* auct. non L.

VIḌANGA

Embelia ribes Burm. f. Euphorbiaceae
Embelia tsjeriam—cottam Dc.
=*E. robusta* C.B. Clarke (non Roxb.)

VIDARI

Pueraria tulerosum Dc. Fabaceae

Ipomoea mauritiana Jacq.
=*I. paniculatus* (L.) R. Br.
=*I digitata* auct. Plur non L.

VIṢAMUṢṬI—KUPĪLU

Strychnos nuxvomica Linn. Ranunculaceae

VIRATARU

Dichrostachys cinerea W. & A. Mimosaceae

VṚKSĀMLA

Garcinia indica Chois. Guttiferae

VṚSCIKĀLĪ

Urtica diocia Linn. Urticaceae
Girardinia diversifolia (Link) Fries.
=G. heterophylla Decne.
=*G. heterophylla* var. palmata Hook. f.
=*G. heterophylla* var. *Zeylanica* (Decne.) Hook. f.
Pergularia daemia (Forsk.) Chiov.
=*Asclepias daemia* Forsk.
=*Daemia extensa* R. Br.
=*P. extensa* R. Br.
Tragia hispida Will.
=*Tragia involucrata* var. *angustifolia* Hook. f. Pax & Hoffmann.
Tragia muelleriana var. *cordatus* (Muell.-Arg)
=*T. involucrata* var. *cordatus* Muell-Arg
Tragia plunkenetii A.R. Smith.
=*T. cannabina* L. f.
=*T. involucrata* L. var. *Cannalina* Hook. f.
Urtica ardens Link.
=*U. parviflora* Roxb.
U. mairei H. Lev.
=*U. parviflora* auct. non Roxb.
U. pilulifiera Linn.

VṚSCĪRA

PUNARNAVĀ *Boerhavia diffusa* L. Nyctaginaceae

VARSABHU *Trianthema decandra* Linn. Ficoidaceae

VETASA

Salix caprea Linn. Salicaceae

JALAVETASA *Salix tetrasperma* Roxb.

VETRA

Calamus tenuis Roxb Palmae

Calamus rheedi Griff.

C. rotang Linn.

C. travancoricus Bedd. ex Hook. f.

Daemonorops draco Blume.

=*C. draco* Will.

VRIHI–ŚĀLI–ṢAṢṬIKA

Oryza sativa Linn. Poaceae

VANAHARIDRĀ

Curcuma aromatica Salisb. Zingiberaceae

VAJRADANTĪ

Potentilla fulgens Wall ex Hook.

=*P. fulgens* var. *intermedia* Hook. f.

P. gerardiana Lindl.

=*R. fragarioides* sensu HK. f. auct. non L.

P. fruticosa L. var. *rigida* (Wall ex Lechm.)

=*P. rigida* Wall. ex Lehm.

=*P. fruticosa* auct. non. L., Hook. f. wolf.

=*P. nepalense* D. Don.

P. fruticosa L. var. *pumila* Hook. f.

P. sundaica (Bl.) Kuntze.

=*Fragaria sundaica* Bl.

=*P. kleiniana* W. & A.

=*P. anemonefolia* var. *kleiniana* (W. & A.) Kitagawa.

=*P. nepalensis* Hook. f.

=*P. reptans* Linn.
=*P. supina* Linn.
=*P. atrosanguinea* Lodd.

VĀSANTĪ

Jasminum arborescens Roxb. Oleaceac

VĀRṢIKĪ

Jasminum sambac Ait Oleaceae

VṚDDHADĀRU

=*Argyreia nervosa* (Burm. f.) Boj. Convolvulaceae
Argyreia nervosa Sweet.
Operculina petaloidea (choisy) Ostrtroom.
=*Ipomeca petaloidea* Choisy.
=*Merremia crispatula* Prain.
Ipomoea pes-caprne (L,) R. Br. ssp. *pes-caprae* Ost.
=*I. biloba* Forsk.
Ipomoea pes-caprae ssp. *brasillensis* (L) Ooststroom.
=*Convolvulus lilobus* Roxb.
=*I. biloba* Forsk.

VĀRIPARNI–JALKUMBHI

Pistia stratiotes Linn. Araceae
Eichhornia crassipes Solms.

VIṢNUKANDA

Crinum defixum Ker.-Gawl. Amaryllidaceae

VANAPSIKĀ

Viola odorata Linn Violaceae
V. cinerea (Boiss.) var. *stockü* (Boiss.) W. Beck.
=*V. stocksü* Boiss.
=*V. cinerea* auct. non Boiss.
=*V. cinerea* ssp. *kathiawarensis* W. Beck.
V. canesens wall.
=*V. serpens* wall var. *canesens* (wall.) Hook. f. & Thoms.
V. biflora L.
=*V. manaslensis* Mackawa.

V. wallichiana Ging ex Dc.
=*V. reniformis* wall.
=*V. biflora* auct. non. L.
V. pilosa Bl.
=*V. serdens* wall ex Ging.
=*V. serpens* var. *glabra* Hook. f. & Thoms.
=*V. palmaris* Buch.-Ham.
=*V. betonicifolia* J.E. Smith.
=*V. patrinii* sensu Hook. f. & Thoms.
=*V.* inconspicua auct. non Bl.
=*V. betonicifolia* Smith. ssp. *nepaulensis* (Dc) Becker.
V. betonicifolia Smith. ssp. *jaunsariensis* (Beeker) Nara
=*V. patrini* var. *suaveolens* watt.
=*V. kashmiriana* Becker.

VANDHYĀKARKOṬAKI

Momordica diocia Roxb. — *Cucurbitaceae*
M cochinchinensis Spreng.

VAJRAKANDA

Sauromatum venuosum (Ait.) Kunth. — Araceae
=*Arum venosum* Ait.
=*S. pedatum* (willd.) Schottt.
=*S. guttatum* (walt.) Schott.
S. brevipes (Hook. f. N.) E. Brown.
=*Tiyphonium brevipes* Hook. f.

VYĀGHRAIRANḌA

Jatropha curcas Linn. — Euphorbiaceae
Jatropha glandulifera Roxb.
J. gossypifolia Linn.
J. multifida Linn
J. nana Delz. & Gibs.
J, integerrima Jacq.
=J. *pandurifolia* Andr.
J. integerrima var. *coccinea* (Link.) Balak.
= *J. pandurifolia* var. *coccinea* (Link.) Pas.
J. regia L. var. Kumaonia C. Dc.
=*J. regia* auct. non L.

ŚANKHAPUṢPĪ

Convolvus microphyllus Sieb ex spreng. Convolvulaceae
=*C. pluricaulis* Choisy.
Evolvulus alsinoides Linn.
Canseora decussata Schult.
Clitoria ternatea Linn.

ŚĀṬĪ

Hedychium spicatum Ham. ex Smith. Zingiberaceae

ŚAṆA

Crotalaria juncea Linn. Fabaceae
C. pallida Ait.
=*C.* mucronata Desv.

ŚAṆAPUṢPĪ

Crotalaria verrucosa Linn.
Crotalariā retusa Linn. Fabaceae
C. spectabilis Roth.
=*C. sericea* Retz.
=*C. leschenultü* Dc.

ŚATAPUṢPĀ

Anethum graveebns Linn. Apiaceae

ŚATÂHĀ

=*Peucedanum graveolens* (L.) Benth. & Hook. f.
Anethum sowa Kurz.

MĪṢREYĀ

Foeniculum vulgare Mill.

ŚATĀVARĪ

Asparagus racomosus willd. Liliaceae
A. curillus Buch.-Ham. ex R*oxb.*
=*A. nepalensis* Baker.
Asparagus fillicinus Buch.-Ham.
A. sarmentosus Linn.

SAMI

Prosopis cineraria (L.) Druce. Caesalpiniaceae
=*P. spicigera* Linn.

ŚARA

Saccharum munja Roxb. Poaceae

ŚARAPUṄKHĀ

Tephrosia purpurea (L.) Pers. Fabaceae
=*T. hamiltonii* Drumm.
Tephrosia wallichii Graham ex Fawectt & Rendle.
=*T. purpurea* sensu Baker.
T. pumila (Lamk.) Pers.
=*T. purpurea* var. *pumila* (Lomk.) Baker.
T. maxima (L.) Pess.
=*T. purpurea* var. *maxime* (L.) Baker.
T. villosa Nees.
T. petrosa Blatter & Halle.
T. candida (Roxb) Dc.

ŚALLAKĪ

Boswellia serrata Roxb. Burseraceae
=*B. serrata* var. *glabra* (Roxb.) Bennett.
=*B. glabra* Roxb.

SĀKA

Tectona Linn. Verbenaceae

SĀKHOṬAKA

Streblus asper Laur. Moraceae

SĀLA

Shorea robusta Gaertn. Dipterocarpaceae

SĀLAPARNI

Desmodium gangeticum (L.) Dc. Fabaceae
=*D. gangeticum* var. *maculatum* (L.) Baker.
Desmodium veiutinum (Wild.) Dc.
=*D. latifolium* Dc.

=*D. lasiocarpum* Dc.
D. pulchellum Benth. ex Baker.
D. laxiflorum Dc.

SĀLMALI

Bombax ceiba L. Bombaceae
=*B. malabaricum* Dc.
=*Salmalia malabarica* (Dc.) Schott. & Endl.

ŚIṄŚIPĀ

Dalbergia sisoo Roxb. Fabaceae
Dalbergia lanceolaria Linn. f.
D. latifolia Roxb.
D. sissoides Graham ex W. & A.
=*Derris latifolia* var. *sissoides* (W. & A.) Baker
D. pinnata (Lour.) Prain.
D. volubilis Roxb.
D. sericea
=*D. stenocarpa* Kurz.
D. horrida (Dennst.) Mabberley.
=*D. spinosa* Roxb.
=*D. sympathetica* Nimmo.
=*D. multiflora* Heyne.

ŚIGRU—ŚOBHĀṄJANA

Moringa obifera Lamk. Moringaceae
=*M.* pterygosperma Gaertn.
M. concanensis Ninumo.

ŚITIVĀRA

Celosia argentea Linn. Amaranthaceae

ŚIMBĪ—NIṢPĀVA

Lablab purpureus (L.) *Sweet.*
=*Dolichos lablabilinn.* Fabaceae
=*D. lignossus* L. & D. *benghalensis* Jacq.
=*D. purpureus* L.

ŚIRIṢA

Albizia lebbeck Benth. Mimosaceae

A. odoratissima (L. f.) Benth.
=*A. odoratissima* var. *mollis* Benth ex Baker.
A. procera (Roxb.) Benth.
Albizia amara Bolvin.
A. chinensis (osbeck) Merrill.
A. stipulata (Roxb.) Boiv.
A. chinensis var. *smithiana* (Roxb.) Sahni.
=*A. stipulata* sensu Baker.
A. julibrissin Durazz. ssp. *mollis* (wall.) Ali & Saeeda
=*A. julibirissin* Durazz.
[*Albizia*=*Albizzia*]

ŚUKANĀSĀ

Corallocarpus epigaeus Benth. ex Hook. *f.* Cucurbitaceae

KAṬUNĀHI

Emicostema hyssopifolium (willd.) Verdoora
=*E. littorale* Bl. Syn. *Gentiana verticillata* L.

ŚṚṄGĀṬAKA

Trapa natans L. var. *bispinosa* (Roxb.) Makino. (Roxb) Onagraceae
=*T. bispinosa* Roxb.
T. natans L. var. *quadrispinosa* Makino
=*T. natans* auct. non L.

ŚAILEYA(KA)

Parmelia periata Ach. Parmeliaceae

ŚAIVĀLA

Zannichellia palustris Linn. Zannichelliaceae
Vallisneria spiralis Linn.
Ceratophyllum demersum Linn.
Potamogeton pectinatus Linn.
Gracilaria lichenoides Harv.

ŚYĀMĀKA

Echinochloa frumentacea Link. Poaceae

E. crusgalii (L.) P. Blauv.
=*E. glabrescens* Muntro.

ŚLEṢMĀTAKA

Cordia dichotoma Forst. Boraginaceae
=*C. myxa* sensu cl.
=*C. lowriana* Brandis.
C. myxa L.

Cordia gharaf (Forsk.) Ehrenb.
=*C. rothii* R. & S.
C. obliqua var. *wallichii* (G. Don.) Cl.
=*C. wallichii* G. Don.
=*C. obliqua* zar. *tomentosa* (wall.) Kazmi.

SILĀRASA

Liquidamber orientalis Miller. Hamamelidaceae
Altingia excelsa Noronha.

ŚATAPATRĪ (TARUṆĪ)

Rosa centifolia Linn. Rosaceae

Rosa alha Linn.
R. bankaise R. Br.
R. chinensis Jacq.
=*R. indica* Linn.
R. damascena Mill.
R. gallica Linn.
R. multiflora Thunb. var. *carnea* Thory.
=*R. muthiflora* auct. non Thunb.
R. cinophylla Thory.
=*R. imvolucrata* Roxb ev Lindl.
R cymosa Tratt.
=*R.* microcarpa Lindl.
R. laevigata Michx.
=*R. simica* auct. non L.
R. maerophylla Lindl.
=*R. macrophylla* var. *hookeriana*. Hook. f.
R. multiflora Thunb.
=*R. multiflora* auct. non Thunb.
R. roxburghii Tratt
=*R. microphylla* Roxb.

ŚITIVĀRA

Marsilea minuta Linn. Rhisocarpae

ŚIVALIṄGĪ

Diplocyclos palonatus var. *walkerū* (chakr.) Bhandari & Singh, Cucurbitaceae
=*Bryonopsis laciniosa* (L.) Naud. var. walkeriū chakr.

ŚATAPATRĪ (SEVANTI)

Dandronthema indicum L. Des. Asterocceae
Chrysanthemum Indicum L.

GULDĀUDĪ

Pyrethrum indicum Linn.
=*Tanacetum cinerarifolium* (Trev.) Sch.
=*C. cinerifolium* (Trev.) Vis.
C. coccinium willd.
C. coronarium Linn.

SATĪNA

Pisum sativum Linn. Fabaceae
P. arvense Linn.

SAPTAPARṆA—SAPTACCHADA

Alsṭonia scholaris R. Br. Apocynaceae
Alstonia venenata R. Br.
A. spectabilis R. Br.

SAPTALĀ—SĀTALĀ

Acacia sinuata (Lour.) Merrill. Mimosaceae
=*A. concinna* (Willd.) Dc.
=*A. rugata* (Lamk.) Merrill.
Euphorbia dracunculoides Lamk.
=*E. angustifolia* Buch.—Ham.
Euphorbia pilosa Linn.

SARALA

Pinus roxburghū Sarg. Pinaceae

=*P. longifolia* Roxb.
P. griffithi M'Clell.
P. gerardiana Wall.
Pinus khasya Royle.
P. markusii Jungh.

SARPAGANDHĀ

Rauwolfia serpentina Benth. ex Kurz, Apocynaceae
Rauwolfia tetraphylla L.
=*R. canescens* L.
=*R. tomentosa* Jacq.
=*R. hirsuta* Jacq.
R. densiflora Benth. ex Hook. f.
R. decurvo Hook. f.
R. micrantha Hook. f.

SARPACCHATRAKA—SARPĀ

Arisaema speciosum (wall.) Mart, var. Aracaceao
mirabile (Sehott.) Hara.
=*A. speciosum* Schott.
=*A. mirabile* Schott.
A. concinnum Schott.
A. tortuosum (Wall.) Schott. var. *curvatum* (Roxb.) Engler.
=*A. curvatum* Roxb.
=*A. tortuosum* sensu Hook. f.
A. tortuosum var. *neglectum* (Schott.) Fischer.
=*A. neglectum* Schott.
A. helleborifolium Schott.
A. laschenaultii Bl.

SARṢAPA

Brassica campestris var. *sarson* Prain.
B. rapa L. ssp. *campestris* (L.) Clapham.
=*B. campestris* L.
B. napus L. ssp. *napus* var. *trilocularis* (Roxb) Schulz,
=*B. rapa* var. *trilocularis* (Roxb) Kitamura.

SĀRIVĀ

ŚVETA

Hemidesmus iudicus R. Br. Asclepiadaceae

KṚSNA

Ichnocarpus frutescens R. Br. Apocynaceae
Cryptolepsis buchanani Roem & Schult.
Tylophora fasciculata Ham. ex Wight.

SIMBITIKĀ

Malus sylvestris Mill. Rosaceae
Pyrus malus Linn.

SŪCĪ

Atropa acuminata Royle. Solanaceae
=*A. belladona* sensu cl., *A. belladona* Linn.

SUNIṢAṆṆAKA

Marsilea minuta Linn. Marsileaceae

SUVARCALĀ—TILAPARṆĪ-AJAGANDHĀ

Cleome gynandra L. Cleomaceae
=*Gyanandropsis pentaphylla* (L.) Dc.
Cleome gynandra L. var. *nana* (Blatt. & Hall) Bhandari.
=*G. pentaphylla* var. *nana* Blatt & Hall.
Cleome icosandra Linn.
=*C. viscosa* Linn.
C. monophylla Linn.
C. vahiliana Farsen.
=*Cleome brachycarpa* vahl ex Dc.
=*C. brachycarpa* var. *longipetiolata* Sabnis.
=*C. brachycarpa* var. *glauca* Blatter & Halib.
C. chelidonni Linn. f.

SAIREYAKA

PĪTA

Barleria prionitis Linn. Acanthaceae

B. prionitis L. ssp. pubiffora (Benth. ex Hohen) Brummitt & Wood.
=*B*. pubiflora Benth. ex Hohen.
=*B*. *prionitis* auct. non L.

RAKTA-ŚVETA

Barleria cristata Linn.
=*B. ciliata* Roxb.
B cristata Linn. var. *dichotoma*

NĪLA-KṚṢNA

B. strigosa willd.
Acanthus ilicifolius Linn.

SVARṆAKṢĪRĪ

KAṬUPARṆĪ

Argemone mexicana Linn. — Papaveraceae
Euphorbia thomsiana Boiss.
Garcinia morella Desr.

SNUHĪ

Euphorbia nerūfolia Linn. — Euphorbiaceae
E. ligularia Roxb.
=*E. nerūfolia* sensu Hook. f.
E. trigona Haw.
E. barnhattū Croizat.
=*E. trigona* sensu Hook. f.
E. antiquorum Linn.
E. royleanā Boiss.
E. tirucalū Linn.
E. pepius Llnn.

SPAṚKKĀ—NIRMĀLYA

Delphinium zalil Aitch. & Hemsl. — Ranuneulaceae
D. brunonianum Royle.
Delphinium denudatum Wall.
D. vestitum Wall ex Royle.
=*D. vestitum* var. *sphenolobum* Bruhl ex Huth.

Schizachrym exile (Hochst.) Stapf.

SOMA (VALLĪ)

Ephedra gerardiana wall. Gnetaceae
=*E. vulgaris* Hook. f. non A. Rich.
E. intermedia Schrenk & Mey.
E. major Host.
=*E. nebrodensis* Tineo.
Sarcostemma acidum (Roxb.) Voigt.
=*Asclepias acida* Roxb.
=*S. brevistigma* W. & A.
Sarcostemma brunonianum W. & A.
S. stocksū Hook. f.
S. secamone (L.) Bennet.
=*S. esculentum* (L.f.) Holm.
=*Oxystelma esculentum* (L.f.) R. Br. ex Schutt.
=*O. secamone* (L.) Karst.

SARPĀKṢĪ

Ophiorrhiza mungas Linn. Rubiaceae
Ophiorrhiza repens (Wall. ex G. Den.) Bennet.
= *Mussaendra repens* Wall. ex G. Don.
= *O. calcarata* Hook. of.
O. rugora wall.
=*O. harr'siana* Heyne var. *rugosa* (wall.) Hook of,
Polygonum plebejum R. Br.

SUDARŚANA

Crinum latifolium Linn. Amaryllidaceae
= *C. zeylanicum* Linn.
C. asiaticum Linn.
C. defixum Ker.-Gawl.

STHALAKAMALA

Hybanthus enneaspermus (L.) F. Muell. Violaeeae
=*Viola enneasperma* L.
=*V. suffruticosa* L.
=*Ionidium suffruticosum* (L.) R. & S.

=*I. enneaspermum* (L.) Vent.
=*I. suffruticosum* Ging.
Hybanthus travancoricus (Bedd.) Melch.
=*I. travancoricum* Bedd.
Hibiscus mutabilis Linn.

SINDŪRĪ

Bixa oreliana Linn. Bixaceae

SAUVĪRABADARA

Zizyphus sativa Gaertn. Rhamnaceae

SADĀPUṢPĪ

Catharanthus roseus (L.) G. Don.
=*Vinca raesa* (L.) Reichhb.
C. pusillus (Murr.) G. Don
=*V.* pusilla Merr.

SĪTĀPHALA

Annona squamosa Linn. Anonaceae

SAPTARAṄGĪ

Salacia chinensis L. Calastraceae
=*S. primoides* Dc.
=*S. latifolia* Wall ex Lawson.

SIDDHAKA

Lagerstroemia speciosa (L.) Pers. Leythraceae
=*L. flos-reginae* Retz.
L. indica Linn.

SAÑJĪVANĪ

Selaginella rupestris Spreng. Sellaginellaceae

STHOUṆEYAKA

Taxus baccata Linn. Taxaceae

SURYABHAKTĀ–SŪRYAMUKHĪ

Helianihus annus Linn. Asteraceae

SARPĀ

Arisaema jacqucmontū Araceae

HIṄGU

Ferula foetida Regel. Apiaceae

Ferula narthex Boiss.
F. alliacea Boiss.
F. galbaniflua Boiss ex Buhse.
F. sumbui Hook. f.

HIṄGUPATRIKĀ

F. jaeschkeana Vatka.
F. narthex Boiss.

NĀḌĪHIṄGU

Gardenia gummifera Linn. f.

HRĪVERA—BĀLAKA

Pavouia odorata willd. Malvaceae
Coleus vettiveroides K. C. Jacob.

HIJJALA

Barringtonia acutangula (L.) Gaertn. ssp.
spicata (Bl.) Payens. ecythidaceae
B. asiatica (L.) Kurz.
=*Mammea asiatica* L.
=*B. speciosa* J. R. & G. Forst.

HASTIKARNA

Leea macrophylla Roxb. ex Horn. Vitaceae
=*L. integrifolia* Roxb.
=*L. robusta* Roxb.
=L. *cinerea* Lawson.
=*L. coriacca* Lawson.
=*L. angustifolia* Lawson.
=*L. parallela* Wall ex Lawson.
=*L. latifolia* Wall ex Kurz.

=*L. tolbottū* King ex Talbot.
=*L. venkobarowū* Gamble.

HṚTPATRĪ

Digitalis purpurea Linn. Scrophulariaceae
D. purpurea Linn.

HAṀSAPADĪ—HAṀARĀJA

Adiantum lunulatum Burn. Polypodiaceae
A. cappilus-veneris Linn.
A. venustum G. Don.

HAPUṢĀ

Juniperus communis Linn. Cuppresaceae
J. macropoda Boiss.

HARIDRĀ

Curcuma longa Linn.
=*C. domestica* (Medik) Val.

HARĪTAKĪ—ABHAYĀ-PATHYĀ

Terminalia chebula Retz. Combretaceae
Terminalia citrina (Gaertn.) Roxb.
=*T. citrina* var. *malayana* Cl.

HIṀSRĀ—KANTHĀRĪ

Capparis sepiaria L. Capparidaceae
=*C. sepiaria* var. *vulgaris* Hook. f. & Thoms.
=*C. sepiaria* var. *incanescens* (Dc.) Hook. J. & Thoms.
=*C. sepiaria* var. *retusella* (Thw.) Thw.
=*C. glauca* Wall ex Hook. J. & Thoms.

GṚDHRANAKHĪ

Capparis zeylanica L.
=*C, horrida* L. f.
=*C. hastigera* Hance.
=*C. xanthophylla* Coll. & Hemsl.
C. brevispina Dc.
=*C. zeylanica* sensu Hook. f. & Thoms.

HILAMOCIKĀ

Enhydra fluctans Lour. Asteraceae

HASTIṢUṆḌĪ

Heliotropium indicum Linn. Boraginaceae

HASTIKANDA

Ampelocissus latifolia (Roxb.) Planch. Vitaceae
= *Vitis latifolio* Roxb.
= *V. mantana* Lawson.
Ampelocissus tomentosa (Roth.) Planchon.
= *Vitis tomentosa* Heyne.
A. araneosa (Dalz. & Gibs.) Planch.
= *V. araneous* (D. & G.) Lawson.
A. arnottinna Planch.
= *V. indica* sensu Lawson.

SECTION II

AMROOD

Psidium guajava Linn. Myrtaceae
P. cattleianum Sabine.

ASARON

Asarum europaeem Linn. Aristolochiaceae

ATRILAL

Ammi majus Apiaceae
=*A. visnaga* Lam.
=*Anthriscus cerefolium* Hoffm.

EBENOS

Diospyros ebenum Koenig. Ebenaceae

AMPEACH

Clausena lancium (Lour.) Skeels. Rutaceae
=*Quinaria lansium* Lour.
=*C. wampi* Blanco.
=*C. wampi* Oliv.
C. pentaphylla (Roxb.) Dc.
C. excavata Burm. f.
C. dentata (Willd.) Roem.
=*C. wilddenovü* W. & A.
C. dentata var. nana (Roxb.) Balakr.
C. dentata var. *pubescens* (W. & A.) Balakr.

AYAPANA

Eupatorium triplinerve Vahl. Asteraceae

=*E. ayapana* Vent.
E. cannabinum L. ssp, *asiaticum* var. *heterophyllnm* (Dc.) Kitam.
=*E. heterophyllum* Dc.
=*E. cannnbinum* auct. non L.
E. chinense L.
=*E. reevesū* Wall ex Dc.

INBUDBUD

Artostaphylos uvae ursi Spreng. Ericaceae

ANJUBAR

Polygonum bistoria Linn. Polygonaceae
Polygonum viviparum Linn.

ANANAS

Ananas comosus Merr. Bromeliaceae
=*A. sativus* Schult. f.

ANISUN

Anisi fructus Apiaceae
Pimpinella anisum Linn.
P. diversifolia Dc.
P. acuminata (Edgcw.) C.B. Clarke.
P. saxifrage Linn. var. *dlsectifolia* C.B. Clarke.
P. stocksū Boiss.
P. walliehiana (Miq.) K.N. Gandhi.
P. candolleana W. & A.
P. heyneana Wall.

EPITHUMON

Cuscuta europea Linn. Convolvulaceae
=*C. epithymum* Linn.
Cuscuta hyalina Roth.
C. chinensis Lam.
C. planiflora Tenore.

AFSANTIN

Artemisa absinthum Linn.

Artemisia indiea Willd.

ALU

Solanum tuberosum Linn. Solanaceae

ASA

Myrtus communis Linn. Myrtaceae

UBAISRAN

Rosmarinus officinalis Linn. am iaceae

NAKHUNA

Medlolotus officinalis Willd. Fabaceae
=*Trigonella uncata* Boiss.
M. alba Lam.
M. indica (L.) All.
=*M. parviflora* Desf.

ISKAPECA

Q. pennata Bojer. Convolvulaceae
=*Ipomoea quamoclit* Linn.
=*Quamoclit vulgaris* Chois

UTANGAN

Blepharis edulis Pers. Acanthaceae
B. maderaspatensis (L.) Roth. ssp.
maderaspatensis var. *maderaspatcnsis* Nappar.
=*B. boerhaovifolia* Pers.
B. linariaefolia Pers.
=*B. sindiea* T. Anders.

USHAK

Dorema ammoniacum G. Don. Apiaeeae
Ferula tingitana Linn.

USKURDIYUN

Tencrium scordium Linn. amiaceae

USTUKHUDDUS

Lavandula stoechas Linn. amiaeeae

L. bippinnata O. Ktze.
=*L. burmani* Benth.

UDSALIB

Paeonia emodi Wall.
P. officinalis Linn.
P. emodi Royle fa. *glabrata* (Hook. f. & Thoms.) Hara.
=*P. emodi* var. *glabrata* Hook. f & Thoms.

KENTAURION

Dianthus anatolicus Boiss. Caryophyllaceae
D. caryophyllus Linn.
D. chinensis Linn.
Erythraea centaurium Pers.
Centaurium officinalis.

KATAD-KATIRA

Astragalus heratensis Bunge.
A. strobilifera Royle.
A. gummifera Labill.
A. tribuloides Delile.
A. multiccpa Benth.
=*A leptocentrus* Bunge.
Sterculia urcns Roxb.
Cochleospermum religiosum (L.) Alston.
=*C. gossypium* Dc.
Sterculia rubiginosa Vent.
=*S. balanghas* L. var. *angustifolia*
(Roxb. ex Jack.) Masters.
S. hamiltonū (Kuntze.) Adelb.
=*S. indica* Merr.

KANOUCHA

Phyllanthes madraspatensis Linn. Euphorbiaceae
Salvia spinosa Linn.

KABAR

Capparis spinosa Linn. var. Capparidaceae
spinosa Jacobs.

=*C. spinosa* var. *vulgaris* Hook. f. & Thoms.
C. spinosa L. var. *himalayensis* (Jafri.) Jaeobs.
=*C.* leucophylla sensu collett. non Dc.

KABABCHINI

Piper cubeba Linn. f.
P. canium Rumph.

KHAMAIDRUS

Tcuerium chamaedrys Linn. Lamiaceae

KHAMAIPITUS

Teucrium chamaepitys
Ajuga ehamaepitys Schreber.

KAYPUTI (CAJUPUTI)

Meloleuca leueadendron Linn. Myrtaceae

CALUMBA

Jateorrhiza palmata Miers. Menispermoceae

KALHARPAT

Laminaria sachharine Lam. Algae

KALGA

A. hypochondriascus Linn. Amaranthaceae
Amaranthus hybrides Linn. var.
hypochondriascus Rob.

KAHAVA

Coffea arabica Linn. Rubiaceae

KAKNAJ

Physalis alkakoengi Linn. Solanaceae
P. angulata Linn.
P. peruviana Linn.

KASHMIRI PATTA

=*S. laureola* (Wall) Hook. Rutaceae

Skimmia anquetillia Taylor & Airya Shaw.

KASNI

Cichorium intybus Linn. Asteracae
C. endivia Linn.

KAHU

Lactuca sativa Linn. Asteraceae
L. serriola Linn.
=*L. scoriola* Linn.

KISAUL HIMAR

Ecballium elatorium A. Rich. Cucurbitaceae

COCA

Erothroxylum coca Linn. Erythroxylaceae
E. truxillense Rusby.
E. moonii Hochr.
=*E. acuminatum* (A.) Walp.
=*E. lucidum* Moon ex Hook. f.
E. cuneatum (Miq.) Kurz.
=*E. burmonicum* Griff.
E. monogynum Roxb.
=*E. tndicum* (Dc.) Bedd.

KHAJAMA

Lavandula vera Dc. Lamiaceae
L. officinalis Choise.
L. bipinnata O. Kuntze.
Anisochilus carnosus (L. f.) Wall.
=*L. carnosa* L. f.

KHATMI–GULKHAIRU

Althoea officicinalis Linn. Malvaceae
Aclea rosea L.
=*Althoea rosea* Linn.
Althoea ludwigii Linn.

KHAMAN SAGIR-KABIR

A. *Sambucus ebulus* Linn. Capriofoliaceae
S, wightiana Wall ex W. & A.
=*S. edulus* auct. non L.

B. *S. nigra* Linn.

S. Javanica Reinw. ex Blume
S. hookeri Rehder.
=*S. javanica* auct non Reinw.

KHARNUB

Cratonia siliqua Linn. L.

KHARBAK

Veratrum album Linn. Liliaceae
V. viride Aiton.

KHABAK

Helleborus niger Linn. Liliaceae

KHUBKALAN

Sissymbrium irio Linn. Cruciferae
S. altissimum Linn.
S. loesalū Linn.
Descurainia sophia (L.) Webb. ex Prantl.
=*S. sophia* L.

KHUNSA

Asphodelus fistulosus Linn. Liliaceae
A. tenuifolius Cav.

KHUBBAJI

Malva sylvestris Linn
=*M. sylvestris* var. *mauritiana* (Dc.) Masters.
M. rotundifolia Linn.
M. parviflora Linn.
M. verticillata Linn.
M. verticillata L. var.
chinensis (Mill.) Hu.
= *M. chinensis* Mill.

=*M. parviflora* auct. non L.

GANDNA

Album ampeloprassum Linn. Liliaceae

A. schoenoprasum Linn.

A. tuberosum Rottl.

=*A. clarkai* Hook. f.

=*A odorum* auct. non L.

GANDHPURI

Gaultheria proeumbens Linn.

G. fragrantissima Wall.

G. hookeri el. & Thoms.

=*G. hookeri* var. *angustifolia* Cl.

G. pyroioides Hook. f & Thoms.

=*G pyrolaefolia* Hook f ex Cl.

G. semi-infra (Cl.) Airy-Shaw.

=*Diplycosia semi infra* Cl.

GARBHPHUL

Anastatica herochuntina Linn. Cruciferae

GHAR

Laurus nobilis Linn. Lauraceae

GARIKUN

Agaricus albus Agaricaceae

Polyporus officinalis Fries Polyporaceae

GULABBAS

Mirabilis jalapa Linn. Nyctaginaceae

GULMEHNDI

Impatiens balsamina Linn. Balsaminaceae

I. chinensis L. var. *supicola* (Hook. f.) Bhaskar.

=*I. rupicola* Hook. f.

I. tripetala Roxb.

GULALALA

Anemone obtusifotia D. Don. Ranunculaceae

=*A. obtusiloba* D. Don. var. *glabra* Hook. f & Thoms.
=*A. multisepala* Qureshi & chaudori.
(Type—N.W.F.P.)
=*A. neelamiona* Q. & C. (Type—Kashmir)
A. marcissiflora Linn.
A. rupestris Hook. of. & Thoms.
ssp. *gelida* (Maxim.) Lauener.
var. *willichū* (Bruhl,) Lauener.
A, trullifolia Hook. f. & Thoms.
var. *linearis* (Bruhi) Hand.

GULABJAMUN

Syzygium jambos Linn. Myrtaeean
=*Engcnia jambos* Linn.
=*Jambosa'vulgaris* Dc.

GONDNI

Briddelia moutana Willd. Euphorbiaceae
B. retusa Spreng.
B. squamosa (Lamk,) Gehrmann.
=*B. rctusa* sensu Hook. f.

GOBHI

Brassica oleracea Linn. var. *botrytis* Linn. Cruciferae
B. oleracea Linn' var. *gongylodes* Linn.

CHANAR

Platanus orientalis L. Platanaceae

CHAY

Camellia sinensis (L.) Kuntze. Theaceae
=*Camellia theifera* Griff.
C. sinensis var, *assamica* (Masters) Kitamura.
=*C. theifera* Griff.
=*C. assamica* (Masters.) W. wight.
C. kissi wall.
=*C. eoduca* Brandis.

CHUKANDAR

Beta vulgaris L. var. *rapa* Dum. Chenopodiaccae

B. vulgaris Linn. var. *benghalensis*

CHOBEHAYAT

Guiacum officinale Linn. Zygophyllaceae

JALAPA

Exogonium purga Benth.
=*Ipomoea purga* Hayne
I. jalapa Schiede.

JAVASHIR

Ferula galban flua Boiss ex Buhse. Apiaceae

JUFA

Hyssopus officinalis Linn. Lamiaceae

JAITUN

Olea europaca Linn. Oleaceae
O. ferruginea Royle.
=*O. cuspidata* Wall ex G. Don.
O dioica Roxb.

RASAN

Inula helenium L. Asteraceae
I. royleana Dc.
I. racemosa L.
I grandiflora L.
I barbata Wall. ex Dc.
=*I. grandiflorā* auct. non Willd.

RATANJOT

Onosma echioides L. Boraginaceae
O. hispidum Wall. ex D. Don.
=*O. cchioides* sensu Cl.
=*O. kashmiricum* Johnston.
Geranium wallichianum D. Don. ex Sw. Geran.
Potentilla nepalensis Hook. of.
Clausena pentaphylla Dc.
Anemone obtusifolia D. Don.
Alkanna tinctoria Tousch.

LABLAB

Hedera helix L. Araliaceae
Colonyction aculeatum House

HUBBULMASAKIN

=*Ipomosa bona-nov* Boj.
var. *grandiflora*

LICHI

Litchi chinensis Sonn. Sapindaceae
=*Nephelium litchi* Camb.
=*L. chinensis* L.

LOQUAT

Eriobotrya japonica Lindl. Rosaceae

VARS

Flemingia grahamiana W.A. Fabaceae
Memecylan tinctorium Melastomataceae
M. umbellutum N. Burman
= *M. edula* Roxb.
M. malabaricum (C.B. Clarkges) Cogn.
M. angestifolium Wight.

JONKMARI

Anagallis arvensis Linn. Primulaceae
A. arvensis L. var. *caerulea* (L.) Gouan.
= *A. arnensis* sensu Hook. f.

JOADA

Teucrium polium Linn. amiaceae

TAMATAR

Lycopersicon esculentum Mill. Solanoceae
=*L. lycopersicon* (L.) Britton.
=*L. lycopersicum* (L.) Farwell.
=*Lycopersicum lycopersicum* (L.) Karsten.
L. esculentum var. *leptophyllum* (Dunal) D'Arcy
=*L. cerasiforme* Dunal.
=*L. esculentum* non Mill.

TURMUS

Lupinus albus Linn.

TODRI

Lepidium iberis Linn. Cruciferae

Cardaria draba (L.) Desv.

=*Lepidium draba* L.

L. latifolium Linn.

L. apetalum willd.

=*L*. ruderale auct. non L.

DUKU

Peucedanum grande C.B. Clarke. Apiaceae

P. sucheri Boiss.

P. āhana Ham.

Kamraj *P. nagpurenra* (cl.) Prain.

=*P. glaucum* var. *negpurense* Cl.

PIYARANGA

Thalictrum foliolosum Dc. Ranunculaceae

T. minus Linn.

=*T. thumbergi* Dc.

T. pedunculatum Edgew.

T. secundum Edgew. var.

hazaricum Riedl.

=*T. hazaricum* Stewart.

T. setulosinerve Hara.

= *T. alpinum* L. var. *acutilobum* Hara.

FANJIYUN

Tussilago farfara Linn. Asteraceae

FRASION

Marrubium vulgare Linn.

Otostegia limbata Hook. f.

O. oucheri Boiss.

M. nigra Cratz.

= *Ballota nigra* Linn.

FASRA

Bryonica diocia Jacq. Cucurbitaceae
B. alba Linn.

FINDAK

Corylus aveliana Linn. Cupuliferace
C. colurna Linn.

FITRASALIYUN

Petroselinum crispum (Mill.) Nym, ex auct. Kew.
=*P. sativum* Hoffm.
=*Carum petroselinum* Benth. & Hook, f.
=*Apium petroselinum* Linn.
Cachrya pabularia (L.) Druce.
=*Prongos pabularia* Lindl.

NARGIS

Narcissus trvette Linn. Amaryllidaceae

DARIAI NARIYAL

Lodoicea seychellarum Labill. Palmeae
=*L. maldivica* Pers,

NILKANTHI

Chrozòphora prostrata Dalz. Euphorbiaceae
=*C. plicata* var. 3--Hook. f.
C. prostraţa var. *parviflora* (Klotz. ex Schweinf) Balak.
=*C. parviflora* K. & S.
=*C. plicata* var. 2—Hook. f.
C. oblongifolia (Delila) A. Juss ex Spreng.
=*C. tinctoria* sensu Hook. f.
=*C. obliqua* sensu Hook. f.
C. rottleri (Gbis.) A. Juss. ex Spreng.
=*C. tinctoria* sensu Hook, f.

PALAKJUHI

Rhinacanthus nasuta (L.) Kurz. Acanthaceae
=*Justicia nasuta* L,
=*R, communis* Nees.

R. nasuta var *montana* (cl.) Balak.
=*R. communis* Nees. var. *montana* Cl.

PATCHOULI

Pogostemon patchouli Pell. Lamiaceae
=*P. heyneanus* Benth.
E. heyneanthus Bencth.
=*P. patchouli* sensu Hook. of. (excl. var. *suavis* f.) non Pellet.
P. parviflorus Benth.
P. pubescens Benth.
=*P. parviflorus* sensu Hook. f.

FILFILSSUDAN

Pimenta officinalis Lindl. Myricaceae
P. racemosa J.W. Moore
=*P. acris* Kostel.

FANJKUST

Cyclamen persicum Miller. Primulaceae
C. europaeum Linn.

BIRANJASIF

Achillea millefolium Linn. Asteraceae
(cultigen var. purple heads)
A. santolina Linn.

BALUT

Quercus leucotrichophora A. Camus. Eagaceae
=*Quercus incāna* Roxb.

BEHMAN (ABYAJ)

Centaurea behen L. Asteraceae
C. calcitrapa Linn.
C. *cyanus* Linn.
C. picris Pall.

BEHMAN (SURKH)

Salvia haemotodas Lamiaceae

BAKLA

Vicia faba Linn. Fabaceae

BADIYAN KHATAI

Illicum verum Hook. f. Magnoliaceae

BABUNA

Matricaria ehamomilla L. Asteraceae
M. parthenium
M. aurea Boiss.
M. lasiacarpa Boiss.

BALANGU

Lallemantia royleana Benth. Lamiaceae
=*Draeocephalum royleanum* Benth.

TUKHMALANGA

Salvia santollacfolia Boiss.
=*S. pumila* Benth.
=*S. aegyptiaca* var. *pumila* (Benth.) Hook. f.
Nepeta elliptica Royle ex Benth.

BINAFULUN

Potentilla tormentilla Neek. Rosaceae

BILLILOTAN

Melissu officinalis Linn. Lamiaceae
M. axillarissa Bukh. f.
=*M. parviflora* Benth.
M. flava Benth.
Nepeta hindostana (Roth.) Haines,
=*N. ruderalis* Buch.—Ham ex Hook. f.

BIHI

Cydonia oblonga Mill Rosaceae
=*C, vulgaris* Pers,
=*C. oblonga* Mill.
=*Pyrus cydonia* Linn.

BUTM

Pistacia terrebinthus Linn. Anacardiaceae

BUJIDAN

Tanacetum umbelliferum. Asteraceae

T. fruticulosum Ledele.
= *T. gracile* Hook. f & Th.
T. longifolium wall. ex Dc.
T. tomentosum Dc.

BULIGALI

Polygala senega Linn. Polygalaceae

P. sibirica L. var. *sibirica* Mukerjee.
P, japonica Houtt.
= *P. sibirica* sensu Bennett.
P. crotolarioides Buch-Ham ex Dc.
= *P. crototarioides* Buch. Ham. ex Dc.
var. glabrescens Haines.
P. sibirica Linn. var. *monopetala* (Camb.) Chodat.

MAKKA

Zea mays Linn. Poaceae

INBUSSALAB

Solanum dulcamara Linn. Solanaceae

MANASIM

Canarium commune L. Burseraceae

RUMIMASTAGI

Pistacia lentiscus Linn. Anacardiaceae

MAJRIGUN

Daphne laureola Linn. Thymelaceae
D. mezccrum Linn.
D. guidum Linn.

D. *oleoidus* Salisb.
D. papyracea Wall. ex Steud.
= *D. cannabina* Wall.

MAHIMAHARAJ

Anamirta coccculus (L,) wight. & Arn. Menispermaceae

=*Menispermum cocculus* L.
=*A. paniculatus* Colebr.

MOUSUMBI

Citrus sinensis (L. Osbeck. Rutaceae
C. aurantium var. *sinensis* L.
=*C. aurantium* var. aurantium proper Hook. f.

C. dulcis Pers.

MUNGFALI

Arachis hypogaea Linn. Labaceae

MANGOSTINE

Garcinia mangostana L. Guttiferae

MAIDALAKRI

Litsea glutinosa (Lour.) Robinson. Lauraceae
=*L. sabifera* Pers.
=*L. chinensis* Lamk.
L. monopetala (Roxb.) Pers.
=*Tetranthera* monopetala Roxb.
=*L. polyantha* Juss.
L. cubeba (Lour.) Pers.
=*L. citrata* Bl.
=L. *kingū* Hook. f.
L. clongata (Nees.) Hook. f.
=*Daphnidium elongatum* Nees.
L. lancifolia Hook. f.
L. zeylanica C. & T. Nees.
L. dccanensis Gamble.
L. tomentosa Heyne ex J. Hooker.
L-floribunda (Blume) Gamble.
=*L.* wightiana sensu J. Hooker.
L. laevigata (Nees.) Gamble.
[*Litsaea—Litsea*]

EUCALYPTUS

Eucalyptus globulus Labill. Myrtaceae
E. citriodora Hook.

SHAKARKANDA

Ipomoea batatas Lamk. Convolvulaceae

SHAKAKUL

Traehydium behmanni Benth. Apiaceae
Eryngium caeruleum Bieb.
E. faetidum Linn.
Polygonatum verticillatum Allioni.

SHALGAM

Brassica-rapa Linn.

SHIRKHIST

Cotoneaster racemiflora Koch. Rosaceae
=*C. nummularia* Fisch.
C. *affinis* indl.
=*C. bacillaris* var. *affinis* (Lindl.) Hook. f.
C. bacillaris wall. ex Lindl.
=*C. affinis* Lindl. var. *bacillaria* (Lindl.) Schneid.
=*C.* affinis Lindl.
C. microphylla wall ex Lindl.
var. *microphylla*
=*C. microphylla* wall. ex Lindl.
Fraxinus ornus L.

SHOUKRAN

Conicum maculatum inn. Apiaceae

SHAILAM

Lolium temulentum inn. Foaceae

SAKMUNIYA

Convolvulus scammonia Linn. Convolvulaceae
C. arvensis Linn.

SAMUNDARSOKH

Salvia plebeia R. Br. Lamiaceae
Argyrea nervosa (Burm. f.) Boj.
=*Argyreu speciosa* Sweet,

=*Convolvulus nervosus* Burm. f.
=*Lettsomia nervasa* Roxb.

SARKHS

Dryopteris filixmas (Linn.) Schott.
=*Aspidium filix-mas* Sw.

SARO

Cupressus sempervirens Linn. Cupresaceae
Casuarina equisetifolia J. R. & G. Forst

SAGUDANA

Cycas pectinata Griff. Cycadaceae
C. rumphii Miq.

SATAR

Zataria multiflora Boiss.
=*Euchla·ma luxuriens* (Durieu) Durieu
=*E. mexicana* var. *luxurians* (Durieu) Haines.
=*Zea mays* ssp. *luxurians* (Durieu) Haines,

SABUNI

Vaccaria pyramidata Medik. Caryophyllaceae
var. *pyramidata*
=*Saponaria vaccaria* Linn.
V. pyramidata var *oxyodonta* (Boiss.) Cullen.
=*Saponaria vaccaria* auct non L.

SASFARAS

Sassfras officinale Nees. auraceae
S. varifolium Kuntze.
S. albidum (Nuttall) Nees.

SISALIYUS

Myr.his odorata Scop, Apiaceae
Meum diffusum

SUMBUL

Ferula sumbul Hook, f. Apiacea

F. grabaniflua Boiss ex Buhse.

SUDAB

Ruta glaveolens Linn. Rutaceae
R. angustifolia Pers.
Euphorbia dracunculoides Lamk.
=*E. angustifolia* Buch, Ham ex D. Don.

SUMAK

Rhus coriaria Linn. Anacardiaceae
R. parviflora Roxb

SURANJAN

TALKH

Colchicum luteum Baker. Liliaceae

SHIRI

Colchicum aitchisenū (Hook. f) Nasir
=*Mereud ra aitchisonū* Hook. f.
=*M. persica* auct. non Boiss & Kotschy.

SOYABEAN

Glycine max (L.) Merrill
Dolichos saja L.
G suja sensu Baker
=*G. soja* Sieb. & Zucc.

HABBULJALAM

Hibiscus cannabinus Linn. Malvaceae

HARMAL

Peganum harmala Linn. Rutaceae

HARSHAF

Cynara scolymus Linn. Asteraceae
Helianthus tuberosus

HALIYUN

Asparagus officinalis Linn Lilaceae

HASHISHATUDDINAR

Humulus lupulus Linn. Moraceae

HASHA

Thymus serphyllum Linn. Lamiaceae

HUFARIKUN

Hypericum perforatum Linn. Hypericaceae

H. japonicum Thunberg.
H. patulum Thunb.
H. mysurense wight. & Arn.
H. monogynum Miller ex L.
=*H. chinense* L.
=*H. salicifolium* Sieb. & Zuec,
H. wightianum wall. ex W. & A.
H. uralum Buch—Ham ex D. Don.
=*H. patulum* sensu Dyer.
=*Norysca urala* (Buch.—Ham. ex D. Don.) K. Koch.

BIBLIOGRAPHY

Anonymous : *The Wealth of India* (Raw Materials), Vols. 1-8, C.S.I.R., New Delhi, 1948-69.

Chopra, R.N. et al : *Glossary of Indian Medicinal Plants*, C.S.I.R., New Delhi, 19'6.

—: *Supplement to Glossary of Indian Medicinal Plants*, C.S I.R., New Delhi. 1969.

—: *Indigenous Drugs of India*, 2nd ed., Calcutta, 1958.

Bailey, L.H. : *Standard Cyclopedia of Horticulture*, Vols. 1-3, New York, 1947.

Bennet, S.S R. : *Name Changes in Flowering Plants of India and Adjacent Regions*, Dehradun, 1987.

Cooke, T. : *Flora of Bombay*, Vols. 1-3, B.S.I., Calcutta, reprint, 1958.

Duthic, J.F. : *Flora of Upper Gangetic Plain*, Vols. 1-2, B.S.I., Calcutta, reprint, 1960.

Haines, H.H. : *Botany of Bihar and Orissa*, Vols. I-III, B.S.I., Calcutta, reprint, 1961.

Hooker, J.D. : *Flora of British India*, vols. 1-7, London, 1987-1972.

Kirtikar, K.R. et. al. : *Indian Medicinal Plants*, vols. 1-4, Allahabad, 1933.

Kanjilal, U.N. et. al. : *Flora of Assam*, vols. 1-5, Govt. of Assam, Shillong, 1935.

Osmasten, A.E. : *Forest Flora of Kumaon*, Allahabad, 1927.

Prain, D. : *Bengal Plants*, B.S.I., vols. 1-2, Calcutta, reprint, 1963.

Maheshwari, J.K. : *The Flora of Delhi*, C.S.I.R., New Delhi, 1963.

Babu, C.R. : *Herbaceous Flora of Dehradun*, C.S.I.R., New Delhi, 1977.

Brandis, D. : *The Forest Flora of North-West and Central Indla*, London, 1874.

—: *Indian Trees*, London, 1931.

Collett, H. : *Flora Simlensis*, Calcutta, 1921.

Hara, H. : *The Flora of Eastern Himalaya*, Tokyo, Japan, 1966-71.

Kihara, H. : *Fauna and Flora of Nepal Himalayas*, Kyota, Japan, 1955.

Raizada, M.B. et. al. : *Flora of Mussorie*, Vol. I, Dehradun, 1978.

Naithani, B.D. ; *Flora of Chamoli*, B.S.I., Allahabad, Vol. I-II, 1984-85.

Sadanha, C.J. et. al. : *Flora of Hassan District, Karnataka India*, New Delhi, 1976.

Rau, M.A. : *High Altitude Flowering Plants of West Himalaya*, Calcutta, 1975.

Gupta, R.K. : *Flora Nainitalensis*, New Delhi, 1968.

Airy Shaw, H.K. : *H Dictionary of the Flowering Plants and Ferns* (J.C. Willis), Cambridge (revised 8th ed.), 1973.

Gamble : *Flora of the Presidency of Madras*, 1915-36.

Nadkarni, K.M. : *The Indian Materia Medica*, Bombay, 19 4.

Watt, G. : *A Dictionary of the Economic Products of India*, 1908.

Sharma, P.V. : *Dravyaguna Vigyan*, Part II, Varanasi, 1956.

Singh, B. et. el. : *Glossary of Vegetable Drugs in Brhattrayi*, Varanasi, 1972.

Chunekar, K.C. et. al. : *Bhavapraksha Nighantu*, Varanasi, 4th ed., 1969.

Acharya, Y.T. : *Dravyaguna Vigyan*, Vol. II, Part II, Bombay, 1950.

Dwivedi, V.N. : *Ayurveda Me Aushadhiyan Ka Vargikarana*, I.A.S.R. Jamnagar, 1966.

Singh, R.S. : *Vanoushadhi Nidarshika*, Lucknow, 1969.

Singh, D.J. : *Yunani Dravyaguna Vigyan*, Part III, Lucknow, 1878.

Pandey, G. : *Vanousadhi Manjusha—Maidani Ksetron Ki Upayogi Vanaspatiyan* (by Prof. Ram Nath Vaidya), Hardwar, 1990.

—: *Puspayurveda—Medicinal Flowers of India and Adjacent Regions*, Delhi, 1992.

Raizada, M.B. : *Supplement to Duthie's Flora of Upper Gangetic Plain*, Dehradun, 1978,

Vaidya, B.G. : *Nighantu Adarsa*, Vol. I & II, Varanasi, 1965-68.